Thinkers for Architects

Series editor: Adam Sharr, Cardiff University, UK

Editorial board

Jonathan A. Hale, University of Nottingham, UK
Hilde Heynen, KU Leuven, Netherlands
David Leatherbarrow, University of Pennsylvania, USA

Architects have often looked to philosophers and theorists from beyond the discipline for design inspiration or in search of a critical framework for practice. This original series offers quick, clear introductions to key thinkers who have written about architecture and whose work can yield insights for designers.

Deleuze and Guattari for Architects
Andrew Ballantyne

Heidegger for Architects
Adam Sharr

Irigaray for Architects
Peg Rawes

Bhabha for Architects
Felipe Hernández

Merleau-Ponty for Architects
Jonathan Hale

Bourdieu for Architects
Helena Webster

Benjamin for Architects
Brian Elliott

THINKERS FOR ARCHITECTS

Bhabha
for
Architects

Felipe Hernández

Routledge
Taylor & Francis Group
LONDON AND NEW YORK

First published 2010
by Routledge
2 Park Square, Milton Park, Abingdon, Oxon OX14 4RN

Simultaneously published in the USA and Canada
by Routledge
270 Madison Ave, New York, NY 10016

Routledge is an imprint of the Taylor & Francis Group, an informa business

© 2010 Felipe Hernández

Typeset in Frutiger and Galliard by Wearset Ltd, Boldon, Tyne and Wear
Printed and bound in Great Britain by TJ International Ltd, Padstow, Cornwall

All rights reserved. No part of this book may be reprinted or reproduced or utilised in any form or by any electronic, mechanical, or other means, now known or hereafter invented, including photocopying and recording, or in any information storage or retrieval system, without permission in writing from the publishers.

British Library Cataloguing in Publication Data
A catalogue record for this book is available from the British Library

Library of Congress Cataloging-in-Publication Data
Hernández, Felipe, 1971-
Bhabha for architects / Felipe Hernández.
p. cm. – (Thinkers for architects ; 4)
Includes bibliographical references and index.
1. Bhabha, Homi K., 1949- 2. Architecture–Philosophy. I. Title.
B5134.B474H47 2010
720.1–dc22
                                                  2009037166

ISBN10: 0-415-47745-X (hbk)
ISBN10: 0-415-47746-8 (pbk)
ISBN10: 0-203-85593-0 (ebk)

ISBN13: 978-0-415-47745-1 (hbk)
ISBN13: 978-0-415-47746-8 (pbk)
ISBN13: 978-0-203-85593-5 (ebk)

Contents

Series Editor's Preface   vii
Illustration Credits   ix
Acknowledgements   xi

1   Introduction   1

    Bhabha's theoretical context   9
    Postcolonial theory   14
    Bhabha and architecture   18
    The plan of this book   21

2   Translation   24

    Walter Benjamin and the 'task of the translator'   26
    Translation in Bhabha's work   30

3   Ambivalence   39

    Postcolonial ambivalence   42
    The ambivalence of architectural history   50

4   Hybridity   58

    Bhabha's hybridity   60
    Critiques of hybridity   73
    Hybridity as form in architecture   77
    Representing non-western architectures   82

5   The Third Space   89

    Theorising the Third Space   90

Spatialising the Third Space 93
Third Space and architecture 96

6 The Pedagogical and the Performative — 99

Cultural difference and the agency of minorities 100
The nation and some ideas on nationalism 104
Bhabha's critique of the nation 108
Questioning duality in the history of colonial cities 117
The performative temporality of contemporary cities 120
Architecture and the performative 123

7 Conclusion — 130

Works Cited — 133
Further Reading — 135
Index — 137

Series Editor's Preface

Adam Sharr

Architects have often looked to thinkers in philosophy and theory for design ideas, or in search of a critical framework for practice. Yet architects and students of architecture can struggle to navigate thinkers' writings. It can be daunting to approach original texts with little appreciation of their contexts. And existing introductions seldom explore architectural material in any detail. This original series offers clear, quick and accurate introductions to key thinkers who have written about architecture. Each book summarises what a thinker has to offer for architects. It locates their architectural thinking in the body of their work, introduces significant books and essays, helps decode terms and provides quick reference for further reading. If you find philosophical and theoretical writing about architecture difficult, or just don't know where to begin, this series will be indispensable.

Books in the *Thinkers for Architects* series come out of architecture. They pursue architectural modes of understanding, aiming to introduce a thinker to an architectural audience. Each thinker has a unique and distinctive ethos, and the structure of each book derives from the character at its focus. The thinkers explored are prodigious writers and any short introduction can only address a fraction of their work. Each author – an architect or an architectural critic – has focused on a selection of a thinker's writings which they judge most relevant to designers and interpreters of architecture. Inevitably, much will be left out. These books will be the first point of reference, rather than the last word, about a particular thinker for architects. It is hoped that they will encourage you to read further; offering an incentive to delve deeper into the original writings of a particular thinker.

The *Thinkers for Architects* series has proved highly successful, expanding now to seven volumes dealing with familiar cultural figures whose writings have influenced architectural designers, critics and commentators in distinctive and important ways. Books explore the work of: Gilles Deleuze and Felix Guattari; Martin Heidegger; Luce Irigaray; Homi Bhabha; and Maurice Merleau-Ponty. Forthcoming titles include *Benjamin for Architects* and *Bourdieu for Architects*. The series continues to expand, addressing an increasingly rich diversity of contemporary thinkers who have something to say to architects.

Adam Sharr is Senior Lecturer at the Welsh School of Architecture, Cardiff University, Principal of Adam Sharr Architects and Editor (with Richard Weston) of *arq: Architectural Research Quarterly* published by Cambridge University Press. He is author of *Heidegger's Hut* (MIT Press, 2006) and *Heidegger for Architects* (Routledge, 2007), and joint editor of *Quality Out of Control: Standards for Measuring Architecture* (Routledge, 2010) and *Primitive: Original Matters in Architecture* (Routledge, 2006).

Illustration Credits

Professor Homi K. Bhabha, page 4
Courtesy of Department of English, Director of the Humanities Center at Harvard University

Reconstruction of a Malay House in Kuala Lumpur, Malaysia, page 78
Photo by Felipe Hernández

Reconstruction of the Temple d'Angkor. Paris Exposition 1931, page 85
Courtesy of Topham Partners LLP

Arcade in the Fort Area, Mumbai, India, page 122
Photo by Rahul Mehrotra

Quinta Monroy (before occupation), Iquique, Chile, page 126
Project and photograph by ELEMENTAL Chile

Quinta Monroy (after occupation), Iquique, Chile, page 127
Project and photograph by ELEMENTAL Chile

Acknowledgements

First, I would like to thank Adam Sharr, for his interest in the project, his support and his useful observations, as well as Georgina Johnson-Cook, from Routledge, who also has been supportive and, above all, patient during the past two years.

I am greatly indebted to Homi K. Bhabha who was kind to listen to my ideas when the project was in the early stages, gave me information and offered me his advice.

My gratitude goes also to Jonathan Harris (University of Liverpool), Peg Rawes (The Bartlett School of Architecture) and Christopher Dell for their insightful comments on drafts.

Other people have contributed indirectly to the realisation of this project. So I must thank Mark Millington, Jane Rendell, Iain Borden, Rahul Mehrotra, Jonathan Hale and Neil Leach for conversations, guidance and support during the past ten years of research on postcolonial and architectural theory.

One more person made this project possible: my wife Lea. I am grateful to her for listening to me, for being a fierce critic and reading drafts many times over. More importantly, I want to thank her for being patient with me when I was away writing and neglected her and our son Sebastian, to whom I am also grateful.

CHAPTER 1

Introduction

Postcolonial theory has had a significant effect on the way we understand intercultural relations today and historically. Since the 1980s, the lexicon of postcolonial theory, the concepts it uses to represent cultures and cultural interaction, have penetrated the rhetoric of contemporary politics, international trade and all areas of academia. Needless to say, postcolonial discourse has also had an effect on architecture. In the past 30 years, the work of thinkers such as Frantz Fanon, Edward Said and Gayatri Spivak has permeated into numerous publications which analyse architectural production around the world, both in previously colonised countries and in western metropolitan centres. It is, however, the work of Homi K. Bhabha which has dominated discussions about postcolonial architecture. The fact that Bhabha employs the concept of 'space', and numerous other architectural analogies, has made his work highly appealing to architects and architectural theorists. However, the political dimension of his work prohibits the facile application of his terminology in the study of specific buildings and cities, or in the broader historicisation and theorisation of architecture. The concepts that Bhabha uses in his writings demonstrate that cultures are complex assemblages made up of multiple elements, histories and subject positions (individuals, social groups, class affiliations, genders and sexual orientations). Hence, when used in architecture they establish a strong link with a wide range of issues outside the limits of such a disciplinary field. For that reason, this book argues that the postcolonial methods of critique used by Bhabha could help to develop further our understanding of architecture and its professional practice.

Before going any further, it is necessary to overcome an initial barrier: the word colonialism (in the title postcolonial). Often, the presence of the word colonialism deters people in the field of architecture because it seems to refer to 'old' buildings. Indeed, the period of colonialism may seem somewhat remote to a generation of people born in the last 30 years of the twentieth century. However, many modern nations were negotiating independence agreements and, even, fighting wars of independence until the late 1970s. Hence, there are many people amongst us who can still tell their own experiences of colonialism and narrate the occurrences of what has been termed by historians 'the end of empire': the years after the Second World War when many colonies gained independence from their European and North American rulers. Some of our grandparents, for example, and many parents of those born before 1980 lived through 'the end of empire'. In the British context, most people are familiar with the case of India and Pakistan, which obtained independence in 1947. This historical event, however, was followed by a succession of others: the Republic of Congo became independent from Belgium in 1960, Algeria gained independence from France in 1962 after an eight-year bloody armed conflict, Equatorial Guinea was granted independence from Spain in 1968 and Mozambique became fully independent from Portugal in 1975. I have selected these examples because they represent the four larger colonial European empires – Britain, France, Portugal and Spain – from which I have taken most of the examples in this book. Interestingly, the British soap opera *Coronation Street* had already been launched in Britain, Ozzy Osborne was already singing with the band Black Sabbath and Le Corbusier had already passed away when some countries were in the process of decolonisation. It is not my intention to trivialise the historical importance of decolonisation but to establish a link between the period of colonialism and the present by highlighting events and people that are familiar to a contemporary architectural audience in order to prove that the period of colonialism is not quite as historically distant as it may seem at first sight.

The historical proximity of colonialism explains why many of the strategies used to construct and exercise colonial authority are still employed today and, indeed, influence many issues we accept unquestioningly, for example the notions of democracy, liberalism, equality and cultural diversity. Often we take those concepts for granted and, so, fail to question their adequacy and their implications in the way we live – and study and practise architecture. The unproblematic acceptance of such concepts makes us complicit with the perpetuation of western hegemonic discourses amongst which architecture is included. In other words, colonialism is not an event historically far removed from us, nor is it irrelevant to architecture. On the contrary, as we gradually get to understand the theories and critiques put forward by Bhabha, we will also get to see how architecture coincides with many aspects of colonial authoritarian discourse.

… colonialism is not an event historically far removed from us, nor is it irrelevant to architecture.

Let us now take a biographical detour in order to introduce the life and journey of Homi K. Bhabha. In the preface to the latest edition of his seminal book *The Location of Culture* (2004), Bhabha provides a brief autobiographical account which serves as background for some of the ideas he develops in the rest of the book. This, however, is not the only personal testimony one comes across when reading Bhabha. His work is filled with descriptions of his life, his experience as a migrant and the journey that has taken him from India to the United Kingdom and, then, to the United States where he currently lives and works.

Bhabha tells us that he was born in India but he also stresses that he is a Parsi. The Parsis are a small and relatively unknown minority group, in India and in the rest of the world (although statistics are imprecise, today the Parsi community has approximately 100,000 members worldwide). The history of the Parsis is vast and complex, so I will not elaborate in great detail. However, I will underline a few points which may help to understand some themes in Bhabha's writing and some of its theoretical and personal underpinnings. Amongst these points is the fact that the Parsis arrived in India from Persia. As such, the Parsis are neither Hindu nor Muslim but followers of the prophet Zoroaster, which makes them a religious minority in India. Bhabha argues that the Parsis have been 'hybridised' through the centuries so, today, their rituals pay respect to Hindu customs but simultaneously maintain their own religious and ethnic identity. In Bhabha's view, the Parsi culture is not homogeneous or static but varied and dynamic; it has changed historically as a result of its geographical displacement and due to interaction with other cultures; yet it also remains distinct in many aspects.

Another important issue is the position that the Parsis occupied in the colonial sociopolitical structures during the seventeenth and eighteenth centuries in India. At that time, the Parsis had become a highly educated community familiar with British traditions. They were given administrative positions in the colonial government, received commercial licences and other benefits that helped them to obtain economic prosperity. As a result, the Parsis stood out from other

Indian groups because they were wealthy, ranked highly in the local government and had acquired the cultural traits of the European. In that sense, the Parsis occupied a middle position between larger cultural groups: they were different from the Indian and also from the British, because in spite of the economic wealth they amassed, the Parsis never attained the same status of the British who remained in the position of authority. As will be explained throughout the book, this 'ambivalence' is an issue that haunts Bhabha's writing. In fact, it is one of the most incisive points in his critique of colonialism: the way it constructs colonised subjects through an ambivalent process of simultaneous inclusion and exclusion that places subjects in an intermediate position between the colonised and the coloniser.

Another important facet in Bhabha's life is the time when he was born: only a few years after India became a sovereign nation on 15 August 1947. Although he does not have personal experience of colonialism, he reminisces that his 'childhood was filled with accounts of India's struggle for Independence, its complicated histories of subcontinental cultures caught in that deadly embrace of Imperial power and domination that always produces an uncomfortable residue of enmity and amity' (Bhabha 2004: ix). Referring to his experience in Bombay during his early university years, he says:

> It was lived in that rich cultural mix of languages and lifestyles that most cosmopolitan Indian cities celebrate and perpetuate in their vernacular existence – Bombay Hindustani, 'Parsi' Gujarati, mongrel Marathi, all held in a suspension of Welsh-missionary-accented English peppered with an Anglo-Indian patois that was sometimes cast aside for American slang picked up from the movies or popular music (Bhabha 2004: x).

His description of Bombay highlights a multitude of languages, accents, cultures and nationalities that intermingle but remain apart. The hyphen that Bhabha uses repeatedly helps to represent those identities which coexist but do not mix fully. This rich assortment of peoples of different origins, cultures and languages is what Bhabha highlights and celebrates in his writings. However, it is not a facile celebration but a critical and heavily politicised analysis of the

This rich assortment of peoples of different origins, cultures and languages is what Bhabha highlights and celebrates in his writings.

circumstances in which different sociocultural groups have interacted historically and continue to interact in the space of modern nations; an interaction that is always determined by an uneven distribution of power which produces hierarchical systems of cultural dismissal and racial discrimination.

Arriving at Oxford to study English Literature in the 1970s was the culmination of what Bhabha calls an Indian middle-class trajectory of formal education, an education whose aim is to emulate the canon of elite English taste, traditions and practices. At Oxford Bhabha became an English scholar and a migrant, a highly educated foreigner in the nation of the former colonial master. He entered that intermediate space of irresolution where a person (or an entire social group) can be 'neither one, nor the other'; 'neither here, nor there'. This very condition generated an interest in the work of writers 'who were off-centre; literary texts that had been passed by; themes and topics that had lain dormant or unread in great works of literature' (Bhabha 2004: xi). Bhabha's interest in such writers and literary works has continued to the present and has expanded to the work of female writers, minority artists and film makers whose work is located on the margins, away from the economic thrust that drives cultural production in the world today. Though Bhabha refers continuously to the margins of culture, to areas between cultures – what he calls the Third Space – and, indeed, locates cultural productivity there, by no means is he attempting to glorify the margins and the peripheries. He simply wants to draw attention to the fact that such marginal positions are the most tangible representation of the inequalities that characterise transnational relations in the world today, as well as in the past. That is why he has expressed concern about the impact that a market economy led by western countries has on cultural production on the margins. Speaking about the work of minority artists, and those working on the periphery, Bhabha says:

> I do want to make graphic what it means to survive, to produce, to labour and to create, within a world-system whose major economic impulses and cultural investments are pointed in a direction away from you, your country or your people (Bhabha 2004: xi).

Bhabha's choice of words indicates that he is invested in the statement. The hesitant second person (you), 'your country, your people', generalises in such a way that he himself is part of the generalisation. Bhabha highlights the constraints and difficulties of being part of a minority, ethnic or artistic, working on the periphery against the grain of a market-driven economy. At the same time, Bhabha argues that the work of such marginalised, peripheral artists is politically contentious because it challenges the very structures that have placed them there, on the margins. What is more, the fact that marginal cultural productivity is never fully absorbed by the world-system is a demonstration of both the failures and the limits of democratic representation and of the inability of the world-system fully to eliminate difference (this seemingly complicated point will be explained in Chapter 5).

Bhabha's interest in the cultural products of marginalised, minoritarian peoples working either in the peripheries or invisibly in the centres is amongst the most important reasons why his work is relevant in the field of architecture. The theories advanced by Bhabha provide a sound foundation to develop a critique of the way in which non-western architectures have been inscribed in the history of the field, and to the way they are theorised today in relation to the western canon. I refer here to the fact that buildings produced by non-western architects – or in countries outside Europe and North America – are always historicised and

Bhabha's interest in the cultural products of marginalised, minoritarian peoples working either in the peripheries or invisibly in the centres is amongst the most important reasons why his work is relevant in the field of architecture.

theorised in relation to European architecture. For example, the buildings produced by indigenous people in Africa, Asia and the Americas were considered to be inferior by the European coloniser because they did not correspond with the classical canon. The same has occurred with modern architecture: western historicity serves to reinforce the hegemony of European buildings, architects and discourses on the basis that they precede modernism in other parts of the world (see Chapter 3). The theories that Bhabha employs in his critique of authoritarian structures (colonial or otherwise) are capable of assigning political value to architectures that have been neglected, or undermined, for not complying with dominant architectural narratives; such as those produced by indigenes prior to colonisation and non-western modernism. Bhabha's theories could also assist in the development of more flexible methods of analysis capable of including architectures produced by poor people in slums and squatter settlements in cities around the world, and the alterations that they carry out on the cities where they live. By dominant (or pedagogical) narratives, I refer to academic discourses – like classicism, modernism, deconstruction and, even, architecture itself – which help to support the above-mentioned referential system used to judge architectural production around the world against the European and North American architectural traditions (see Chapter 5).

But let us return to the biographical account with which we started. Bhabha spent most of the 1980s and part of the 1990s in the United Kingdom where he taught at the University of Sussex. During that time he published various articles and essays which truly transformed the study of colonialism and the emergence of postcolonial theory. Some of those essays were compiled in the book *The Location of Culture* (1994) which had an immediate impact on literary, cultural and postcolonial studies. Bhabha's book subscribes to a strong movement which started in the late 1960s and continued throughout the 1980s, that promoted a change in the way that social difference was understood and discrimination was legitimised. Postcolonial critics promulgated a focus on culture and cultural identity, rather than on ethnicity and gender. They turned their attention to the entire set of practices and histories of the people who were formerly derogatorily referred to or classified as *the* Black, *the* Latino, *the* Indian, Women, Gays and so on – banners which homogenise entire sociocultural groups. This is a ground-breaking proposition because it demonstrates that these groups are neither homogeneous, as the banner suggests (*the* Black), nor can they be represented in isolation, because identities are always constructed in relation to one another. In fact, the question of representation lies at the centre of postcolonial discourse because in the process of representing the 'other' (colonised subjects or minorities) as inferior, one is simultaneously representing oneself as superior. These two aspects (the fact that cultures are not homogeneous and that the cultures are inseparably connected, yet not united) are central to the postcolonial critique of authority; the way it is constructed and exercised in colonial and contemporary cultural relations.

Soon after the publication of *The Location of Culture*, Bhabha received a fellowship at Princeton University where he was also Old Dominion Professor. Subsequently, Bhabha took various academic positions at the University of Pennsylvania, Dartmouth College and, in 1997, became Chester D. Tripp Professor in the Humanities at the University of Chicago. Currently, Bhabha is Anne F. Rothenberg Professor of English and American Literature and Language, as well as Director of the Humanities Center at Harvard University. Nonetheless, Bhabha maintains close links with academic institutions in the United Kingdom and India through advisory positions, membership of various boards and steering committees, as well as through lectures and other academic activities in both countries and others around the world.

Bhabha's theoretical context

In general, Bhabha's work has been located in the context of what is known as post-structuralism. The label 'post-structuralism' refers to a group of continental philosophers, mainly French, who criticise the principles of 'structuralism'. According to structuralism there are definite underlying structures that determine the position of subjects in relation to one another. Subjects can be persons or groups of people. Subjects can also be intangible, such as meaning, nationality or culture. The followers of structuralism advocate binary systems of analysis. In such systems one of the components is assigned superiority over the other, and that component determines meaning. One of the most common examples used to explain structuralism is the book, and the inherent binary relationship between the author and the reader. Traditionally, the reader is expected to understand the book, its meaning, its message. Hence, the author is given a dominant position because she/he determines the meaning of the book. The reader, on the other hand, occupies the inferior position, because the reader receives the message, meaning, from the author. Post-structuralism proposes the dismantling of such binary hierarchical structures. An example of such dismantling can be found in the work of Roland Barthes, one of the leading figures of post-structuralism, who wrote an essay entitled 'The Death of the Author'. By metaphorically killing the author, Barthes removes the author from her or his position of authority. Instead of regarding the author as the sole source of meaning of the book, Barthes 'authorises' the reader as an alternative source of meaning. Moreover, since there are many readers, there is also a multiplication of meanings (or interpretations of the book). In sum, with this argument Barthes reverses the hierarchical binary 'structure' between author and reader. What is more, such a reversal can be taken as a basic form of deconstruction which later developed, under the hand of Jacques Derrida, into one of the most significant philosophical currents of the late twentieth century. Chapter 2 of this book examines a similar case of deconstruction in the work of Walter Benjamin and his theory of literary translation. In this case, Benjamin calls into question the primacy of the 'original' over the 'translation'. Bhabha, and other postcolonial theorists, resort to this kind of theoretical operation (deconstruction) in order to revert the authority granted to the culture of the coloniser in relations of colonialism, or to the west in contemporary cultural relations.

Bhabha's work is associated with post-structuralism precisely because it advocates the dismantling of binary systems of social antagonism. In order to dismantle such binary systems, Bhabha employs various concepts: hybridity, the Third Space and cultural difference (a term closely related to the notion of the performative temporality of the nation); all these concepts are carefully analysed in this book. Each one of these terms has different connotations in the overall structure of Bhabha's critique of colonialism and contemporary cultural relations. However, the purpose is always to challenge the reductionism of binary methods of cultural analysis. Cultural hybridity, for example, designates the constant transformations that result from sustained processes of cultural interaction. Hybrid cultures are unclassifiable because they are different from both colonised and coloniser. What is more, since processes of hybridisation are permanent, they cause a proliferation of cultures, *à la* Barthes, which no longer fit within the limitations of binary systems of social antagonism (see Chapter 4). The Third Space, in Bhabha's usage, is a third location, outside, or in-between, traditional binary structures of cultural analysis. The Third Space is an attempt to assign spatial characteristics to the margins, those areas of irresolution between cultures, or inside them, where hybridisation occurs. Again, as a 'third' term, it offsets the theoretical operability of binary structures of analysis (see Chapter 5). Cultural difference is one of Bhabha's most powerful and politically charged concepts. It refers, particularly, to contemporary cultural relations and the conditions in which we live today. Bhabha proposes the concept of cultural difference in order to advance a critique of other notions such as cultural diversity and multiculturalism which are very familiar to us all. These terms indicate the coexistence of different cultures within the space of modern nations. However, as Bhabha points out, the promotion of multiculturalism goes hand in hand with various forms of containment of diversity. This is because foreign peoples (cultures) are allowed within the space of the nation only as long as they comply with the parameters established by the host society for their interaction. Bhabha argues that countries which participate in and promote multiculturalism assert their commitment to diversity under the tacit condition that the demographics of diversity consist of highly skilled, highly educated and economically productive migrants rather than the poor (refugees, political exiles or unskilled labourers). In other words, such buzzwords of contemporary liberal politics, cultural diversity and multiculturalism, belong to a rhetoric of cultural stratification, although they claim the opposite. Another problem inherent in those two terms is that they represent cultures as homogeneous totalities, i.e. *the* British, *the* Indian, *the* Black, *the* Latino. Such a kind of cultural representation erases differences of class, gender, religion and, even, race which are inherent in all of them – for, as mentioned above, it is clear that neither *the*

British, nor *the* Indian, nor *the* Black, nor *the* Latino are, in fact, homogeneous groups. Thus, the purpose of the concept of cultural difference is to unveil the inherent differences that exist in all cultures. As a result, cultural interaction can no longer be approached in terms of a polarity between holistic cultural systems such as the Indian and the British, the colonised and the coloniser or the east and the west. The study of the interaction between cultures has to account, also, for the existence of internal differences. Therefore, cultural difference illustrates a proliferation of difference which distorts the perception of authority as constructed by binary systems of sociocultural antagonism (see Chapter 6).

This brings us to the influence of psychoanalysis, which in Bhabha can also be seen within post-structuralism in the work of Jacques Lacan and Julia Kristeva, both of whom were or are qualified psychoanalysts; although Bhabha is also influenced by Sigmund Freud. Psychoanalysis has been instrumental for Bhabha in order to question the homogeneity and completeness of identity, because in psychoanalysis all identities are incomplete, whether they are individual or collective. According to psychoanalysis, the mind is not a coherent structure that develops linearly in time from childhood to adulthood. On the contrary, psychoanalysts argue that the human mind can be described as being somewhat disjointed and that its development does not follow a linear pattern of chronological development. In their view, the mind is made of, or inhabited by, a multitude of experiences which coexist there in disarray, some are chronologically distant and others are more recent, yet together those experiences determine our behaviour. By the same token, according to psychoanalysis, identities (individual and collective) are always in the process of being formed and that process is neither linear nor cohesive. For that reason, identities are understood as being both plural and dynamic, always changing; even our own individual identities display multiple facets. Bhabha borrows from psychoanalysis terms such as doubling, desire, narcissism and, more importantly, ambivalence. As will be discussed in Chapter 3, psychoanalytic theories have helped Bhabha to re-imagine colonialism – colonial discourse and knowledge – in terms of what he calls an 'ambivalent doubling', the simultaneous identification and disavowal of colonised subjects.

Bhabha has applied his critical methods of reading to other unexpected 'texts' including visual arts, human rights, heritage and, also, architecture.

Considering that Bhabha is, above all, a literary critic, the best way to describe his work is as interdisciplinary. His interdisciplinarity consists of using different methods of analysis and disciplinary 'knowledges' – literature, philosophy, psychoanalysis, sociology, history and so forth – in order to 'read' a multitude of texts: historical, theoretical and legal documents, as well as fiction (chronicles, dramas, novels and poems written in different historical periods). As a literary critic, Bhabha 'reads' in different ways, for example he close reads, reads comparatively or critically amongst other forms of reading. By reading in different ways, he pays attention to various aspects of the text, its technicalities (i.e. grammar), its rhythms, figures, hesitations and so on, in order to advance an interpretation of them. It comes as no surprise that he refers to colonialism, the nation and, even, to modernity as narratives which need to be read and interpreted critically. Bhabha has applied his critical methods of reading to other unexpected 'texts' including visual arts, human rights, heritage and, also, architecture. That is why some commentators describe the work of Bhabha, and postcolonial methods of critique in general, as a 'form of reading'; a way of reading and interpreting history from the perspective of previously colonised peoples and other minorities. As Bill Ashcroft explains, postcolonial reading:

> is a form of deconstructive reading most usually applied to works emanating from the colonisers (but may be applied to works by the colonised) which demonstrates the extent to which the text contradicts its underlying assumptions (civilisation, justice, aesthetics, sensibility, race) and reveals its (often unwitting) colonialist ideologies in the process
> (Ashcroft 1998: 192).

The way in which western texts contradict underlying assumptions such as civilisation and justice will be discussed in Chapters 2, 3 and 4. Suffice it to say for now that the aim of Bhabha's postcolonial reading of historical, literary and legal texts (amongst other genres) is to reveal the contradiction inherent in those texts in such a way that the claim for authority which those texts underpin becomes both questionable and politically unsustainable.

Postcolonial theory

It feels necessary to provide a brief introduction to what is implied in the term 'postcolonial theory', considering that Bhabha is one of its leading exponents. However, this task may prove to be elusive given the fact that, since its inception in the 1980s, postcolonial theory has become exponentially more complex and contentious, especially because it has exceeded the disciplinary boundaries of literature and a few other areas in the humanities to which it was initially confined. As explained above, postcolonial theory refers, most simply, to the rereading of history in order to reveal the economic, cultural, linguistic and social strategies used in order to create and maintain an unequal distribution of power between the colonised and the coloniser. More recently, postcolonial theory and criticism has permeated political and economic debates in order to address not simply the legacies of western colonialism but, more importantly, the inequalities of contemporary international relations. As Bhabha himself puts it:

> Postcolonial criticism bears witness to the unequal and uneven forces of cultural representation involved in the contest for political and social authority within the modern world order. Postcolonial perspectives emerge from the colonial testimony of Third World countries and the discourses of 'minorities' within the geopolitical divisions of East and West, North and South. They intervene in those ideological discourses of modernity that attempt to give hegemonic 'normality' to the uneven development and the differential, often disadvantaged, histories of nations, races, communities, peoples. They formulate their critical revisions around issues of cultural difference, social authority and political discrimination in order to reveal the antagonistic and ambivalent moments within the 'rationalisations' of modernity (Bhabha 1994: 171).

Postcolonial criticism bears witness to the unequal and uneven forces of cultural representation involved in the contest for political and social authority within the modern world order.

It appears that, for Bhabha, the postcolonial is a discourse of the minorities which offers the possibility to question, or challenge, traditional assumptions – attitudes and ideologies – according to which certain social groups are inherently superior to others. To intervene in the ideological discourses of modernity which normalise uneven development and undermine the histories of other nations, races and peoples is Bhabha's way of calling for a thorough revision of the way non-dominant groups have been represented historically and are addressed in the contemporary world order. It is important to understand that Bhabha is not interested in the 'colonial condition' as a thing of the past, but on the effects that colonialism continues to have in the world today. If we look closely, formerly colonised nations – though now independent and sovereign – continue to have less access to global institutions of power; they represent the bulk of the so-called developing world (or Third World). In his critique of contemporary world-order, Bhabha points out that international aid to the poor countries of the developing world does not necessarily lead to the eradication of poverty, nor is the aim of such aid to alleviate the uneven distribution of power in the world. On the contrary, the practices of 'conditionality' with which economic, educational and technological assistance is offered to nations in the developing world guarantee that the First World nations retain their position of power. By conditionality Bhabha refers to practices which require the beneficiaries (the receivers of aid) to give commercial benefits to the lenders. For example, help is given to train people in a developing country so that they can grow a certain crop, money is lent to them so they can build the necessary infrastructure and machinery may also be offered so they can work the fields. In turn, the poor country may be required to sell that product at lower prices to the First World nation that offered the aid and, in addition, the poor country may be asked to offer tax benefits so that the wealthier nation can invest further. Such a way of conducting global government is a continuation of the strategies of domination used to control colonial subjects and to establish authority over them.

The perpetuation of poverty and the subsequent maintenance of global authority on grounds of economic supremacy are not the only problem Bhabha refers to. He is also concerned about the fact that such hierarchical structures validate the dismissal of cultural practices different from those sanctioned by the dominant party. Bhabha often resorts to the world of art as a means to qualify his argument. Modern art, for example, or other forms of western art – 'Brit-Art' comes readily to mind – have dominated the art market for over a century, to the detriment of other practices or practitioners that do not comply with the parameters set by powerful art dealers, galleries and collectors. We already mentioned a similar case in the context of architecture, where non-western architectures are inscribed in the history of the field only when they comply with western scholarly narratives and methods of architectural classification (classic, modern, high-tech, etc.). Ultimately, the aim of postcolonial discourse – at least in Bhabha's terms – is to challenge the discourses that validate such claims for western authority in order to rearticulate, even reformulate, the methods through which knowledge is produced. That way, the products of other peoples and other cultures can be accounted for. In other words, it attempts to open up a space within western academia where it is possible to validate the discourses and cultural products of marginalised peoples.

But what do architects have to do with this? How do these discussions pertain to architecture? The answers are both simple and very complicated, because architecture and architects are largely complicit with the above-mentioned structures of sociocultural and political hegemony. Architecture was one of the principal means used by colonisers to impose a new social and political order and, also, to maintain control over colonised subjects. Colonial cities, for example, served to educate the colonial savage to elevated forms of habitation. Since colonial subjects were considered to be uneducated and backward – in relation to western norms and technology – they had to be taught the way of the European, which included how to live in an orderly fashion in the city (unlike savages). Simultaneously, the rational organisation of buildings (i.e. the orthogonal grid) helped to keep the ethnically different separate, either outside the city walls or on the periphery; away from the city's core inhabited by the colonial elites. For this reason, colonial cities can be considered as the spatial materialisation of the 'civilising mission', while simultaneously representing the violence of colonisation.

Architecture was one of the principal means used by colonisers to impose a new social and political order ...

Another reason why this discussion pertains to architecture, and why architects are largely complicit with colonial and other discourses of domination, is because of the way architectural history is written. As will be demonstrated at various points throughout this book, the history of architecture has been constructed on the basis of a referential system that grants authority to European architecture and, consequently, only inscribes architectures produced by colonised subjects and other minority groups when they correspond with the lineaments of such a referential system. That is why, until very recently (and still today), architectural history books only show non-western buildings when they *look* like western buildings or when they have features that are comparable to them, such as the buildings designed by Balkrishna Doshi in India, Ken Yeang in Malaysia or Oscar Niemeyer in Brazil, to mention only a few. As will be discussed in Chapter 4, the buildings designed by these three architects are commended because their work has attained a great degree of refinement in relation to European norms, or, as William Curtis explains in relation to Doshi, because his 'development epitomises modern architecture adapted to Indian conditions' (Curtis 2000: 572). The point is that Doshi's work is an adaptation of the modern architecture of Le Corbusier, his former employer and mentor. As a result, non-western architectures can only be understood in relation to European norms and aesthetics. In fact, these are the kind of discourses that 'give hegemonic normality to the uneven development and differential histories of nations, races, communities and peoples', that Bhabha refers to in the latter quotation; discourse to which, as we have seen, architects do subscribe.

A third aspect worth mentioning is the fact that such methods of historical inscription dismiss the architectures produced by common people in the act of survival. By this I mean the architectures of poor people in slums, squatter settlements and, also, the appropriations of space that they carry out in the centres of cities in order to live and work, to survive in a world-system that is adverse to poverty. These architectures may not correspond in any way with the referential system used to judge architectural production around the world, but it does respond to the realities and complex needs of minority peoples who live on the margins of culture, between social classes and economic strata and, in many cases, completely outside the axis of global capitalism. That is why Bhabha's own account of postcolonial theory, with its specific socio-political purpose, is relevant in the study of architecture: because it is important to reformulate the methods through which architectural knowledges are produced in order to assign architectural validity to the work of non-western architects and to include the architectures produced by (formerly) colonised subjects, migrants and minority groups in the continuous construction of cities, its spaces and its buildings.

Bhabha and architecture

Bhabha's interest in peripheral literatures and art practices has recently extended to architecture. In 2007, Bhabha was a member of the Master Jury of the Tenth Cycle of the Aga Khan Award for Architecture. This prestigious award was created in 1977 in order to recognise exceptional architectures in which Muslim peoples have had a significant presence. Indeed, this is a complicated task. In principle, such a demarcation creates a polarity between Muslims and non-Muslims. However, it is not the purpose of the award actually to operate on such a reductive premise. 'In serving to enhance architectural innovation in Muslim Societies, the jurisdiction of the Award includes Muslim nation-states, but goes beyond them to recognise Muslim societies as part of the global diaspora of peoples across the world' (Bhabha 2007). With the notion of the 'diaspora' (in this instance referring to Muslim people who live outside Muslim nation-states) Bhabha makes the location of the 'Muslim world' imprecise, unbound by geopolitical borders; his definition includes Muslim nation-states as well as Muslim individuals and communities around the world. It is not that nations and national borders have no relevance in the world today. Bhabha insists that borders do exist and, given the conditions of global market economy, their relevance is perhaps stronger than ever before – nation-states and national borders are needed precisely so that an 'inter-national' economy can exist. However, Bhabha's argument is that, while the nation continues to have political validity, there is an inherent parallel temporality, another dimension in contemporary nations, which transcends geographic borders. This temporality, or alternative dimension of the nation, is embodied in the people of the nation whose mobility in the contemporary world-order escapes facile classification within those rigid boundaries of legislative control (see Chapter 6). It follows that Muslim architecture can no longer be simply associated with specific regions, nor can it be linked to a particular set of formal characteristics – a stereotypical image. In the same way that Bhabha elaborates on 'a shift in vocabulary between Muslim *societies* to Muslim *realities* [which] reflects the way we live today, as part of an inter-cultural, multi-faith world crossing cultural boundaries and national borders' (Bhabha 2007), we must relieve architecture from its dependence on image in order to think of an architecture that responds to the *realities* of the people they are designed for. Here again Bhabha stresses the disparity between our construction of society (Muslim society) usually linked to the idea of a homogeneous community, and the realities of the people of that community; their different ethnicities, religions, social affiliations, political allegiances or simply their location outside the national territories they are intrinsically linked to (i.e. Latinos live in South America, not in Miami or New York).

It is interesting that Bhabha, a non-architect, concludes his report by drawing attention to the 'detail' rather than the edifice (the whole building):

> Scale is a measure of complexity not size, and the detail is often the most expressive element of the whole enterprise. In our search for continuities or differences – across cultures, traditions, urban contexts and rural settings – we too hastily demand broad outlines, stark oppositions, large frameworks and harmonious horizons. Too often, however, the subtle process of cultural

transmission of historical transformation happens at the level of 'moments of overlap' within a larger pattern of representation. Something almost imperceptible emerges at the border or margins of cultural constructions – be it buildings or books – and becomes the harbinger of what is new and innovative (Bhabha 2007).

This quotation has to be read carefully because while Bhabha does allude to architectural details, he is also referring to much broader issues. It seems clear that Bhabha finds great architectural value in the details of buildings, the minuscule components of buildings which refer to the way people actually use them. For Bhabha, architectural details (door knobs, windows, window locks, screens, radiators) convey a sense of humanity, they bring forward the existence of an inhabitant (or inhabitants) and give clues as to the conditions of habitation. We could say, quite safely, that Bhabha is 'reading' buildings, closely, carefully, inquisitively, in the same way he examines literary texts. He is looking for the interplay and juxtaposition of different rhetorical forms, grammar constructions while, at the same time, considering the whole piece; the book, the building. Bhabha's statement also sheds light on a more problematic issue: the fact that architects often look at the 'wrong scale' when tackling the issue of architectural identity. As mentioned above, the historicisation of non-western architectures has been done on the basis of a referential system of architectural production which classifies buildings according to their image, materials or construction techniques. Only rarely is attention given to the way buildings respond to people – not to the environment, but to the individual or collective user(s). That is why Bhabha reminds us that 'in a world of increasing transnational traffic, the signature of specificity and locality – the productive signs of difference – often inhere in the *telling* detail that provides a narrative of the dialogue between tradition and

We could say, quite safely, that Bhabha is 'reading' buildings, closely, carefully, inquisitively, in the same way he examines literary texts.

social change' (Bhabha 2007). Inadvertently, perhaps, or quite deliberately Bhabha's proposition urges that we examine the way we theorise and historicise architectural production in the current conditions of cultural interrelatedness. Bhabha turns the building into a text, a narrative, and so it is connected with both its authors and its readers. At the same time, as he does with culture in general, Bhabha conceives of architecture – buildings, cities, spaces – as a subject in the 'Third Space', that area where culture is at its most productive, because buildings (and cities) are always metaphorically in the middle between architects' interests, developers' economic expectations and planning laws, while also being continually re-signified by users. Certainly, buildings provide the physical spaces where people perform and negotiate their differences. Even though buildings are inert, they are not culturally static; they express those narratives of conflict between peoples (users), power, technology and social change.

Even though buildings are inert, they are not culturally static; they express those narratives of conflict between peoples (users), power, technology and social change.

The plan of this book

This volume of the *Thinkers for Architects* series reveals how the work of Homi K. Bhabha can contribute to develop architectural studies in general, not only in colonial, or formerly colonised parts of the world, but also in the west, not just in the past, but also in the present. To do so, the primary focus is on Bhabha's most influential book, *The Location of Culture*, although essays published in other books, journals and magazines will be brought into the discussion. Chapters 2 and 3 are, to some extent, a continuation of the Introduction. They provide basic historical background about colonialism, the colonising mission and the strategies used in order to construct and exercise colonial power. The terminology of postcolonial theory, which is often considered difficult and confusing, is also explained in these two chapters as a way to facilitate the understanding of other aspects of Bhabha's theory studied further in the book. The main purpose of Chapters 2 and 3 is to introduce the concepts of 'translation' and 'ambivalence' which are essential in order to understand Bhabha's critique and his method of analysis. Chapter 3 contains a discussion about the way in which non-western architectures have been 'ambivalently' inscribed in the conspicuously singular history of architecture. It is a postcolonial reading of architectural history. The purpose of this discussion is to demonstrate, early on, the applicability of Bhabha's method of critique to the study of architecture.

The following three chapters focus on three of Bhabha's most incisive concepts: hybridity, Third Space and the pedagogical and the performative respectively. The tone and style of prose may change slightly as the discussion becomes more theoretical. Do not be dismayed. To ease what may seem complicated, the chapters include commentary of some of the critiques of Bhabha's work. Rather than attempting to dismiss Bhabha's theories, the critiques of his work are brought into the discussion as a way to facilitate its understanding in a negative fashion. In other words, the critiques serve as warning in the sense that they point out the shortcomings of Bhabha's work and, also, tell how 'not' to use his ideas in the context of architecture. The three chapters conclude with extended discussions about architecture. These are examples of the potential inherent in the work of Bhabha to undertake alternative 'readings' of architecture; its practice, theorisation and historicisation. It is also worth noting that the three chapters develop chronologically. Chapter 4 puts an emphasis on the study of colonial architectures, while Chapters 5 and 6 examine contemporary case studies. The purpose of this chronological progression is to show that the postcolonial methods of critique that Bhabha puts forward in his writings can be employed in the study of historical architectures as well as in analysis of contemporary practices.

Chapter 4 focuses on the notions of hybridity and hybridisation, two terms which have been readily appropriated by architects in order to study architectural production outside Europe and North America – although more recently they have also been used to discuss Euro-American architectures. The notions of hybridity and hybridisation are helpful to create alternative methods of architectural analysis connecting the study of buildings and cities with a much wider range of cultural, political and social aspects that are inherent in their production. Chapter 5 discusses another term associated with the work of Bhabha, one which appeals to architects and those involved in spatial studies: the Third Space. Although Bhabha himself does not develop a theory of the Third Space – he uses the term allegorically – other theorists like Henri Lefebvre and Edward Soja have used the idea of the Third Space to advance innovative methodologies of spatial and territorial analysis. The final chapter looks at the 'pedagogical' and the 'performative', two concepts that Bhabha uses in order to develop a critique of the nation as a modern construct based on the principles of the Enlightenment, i.e. rationalism, progress and homogeneity. By disclosing the multiple temporalities of the nation (the pedagogical and the performative) Bhabha brings to the fore the agency of people as the bearers of national culture. Like his other terms – hybridity and the Third Space – the purpose of Bhabha's critique is to locate culture in a liminal zone, between cultural structures. The chapter ends with the analysis of three contemporary case studies in Singapore, India and Chile. The structure of the book has been planned to cover various historical periods and multiple regions of the world, including some regions which have been absent from the bulk of postcolonial architectural theory, as is the case of Latin America. There is also a balance between explicative discussions about the work of Bhabha and the analysis of architectural case studies; the latter highlighting how his arguments offer an opportunity to question the hegemonic normality assigned to certain architectural postures which perpetuate the uneven and differential historical inscription and theorisation of non-western and minoritarian architectures.

CHAPTER 2

Translation

Literary translation confronts the translator with the presence of the original, the work that needs to be translated into another language so that people in other countries, cultures, can have access to it, read it and interpret it. As such, translation implies a displacement across boundaries between languages. Yet, considering that meaning is constructed grammatically and historically, such a displacement is not a simple process. Added to the equation is the fact that meaning is also produced by users when they speak daily, unrestrained either by rules (grammar) or by history. Can meaning be carried forward entirely from one language to another? Will the book be interpreted differently by a readership in another language, culture? Will the book retain its significance as a symbol of 'national culture' or will it be estranged and transformed into a sign of difference? These questions lie at the centre of discussions about literary translation but have implications beyond the margins of that disciplinary field. Indeed, the concept of translation offers ample opportunities to study how non-literary subjects, like architecture, travel across cultural borders (for example, buildings, forms, technologies and the ideas behind them).

Indeed, the concept of translation offers ample opportunities to study how non-literary subjects, like architecture, travel across cultural borders.

On a basic level, colonialism involves various forms of translation which occur more or less simultaneously. For example, the physical translation (displacement) of people to the colony – or vice versa – the subsequent need to communicate in different languages (which requires linguistic translation) and the imposition, in most cases, of religious, economic and political systems which need to be 'translated' (adapted) in order to operate in locations different from Europe. Needless to say, architecture and urban planning also play a part in this process, because architecture can be used both as a symbol of cultural superiority and as a means of socio-political control. How can architecture act in such ways? The answer is relatively simple: in the same way that Christianity is taught as a means to impart civilisation, the coloniser also delivers its own living standards through architecture by offering buildings which are implicitly better than the primitive huts where colonised subjects live. Additionally, the coloniser administers a certain way of organising the territory (planning) in order to make cities work efficiently. So, not only is a mode of habitation imposed on colonised subjects, a whole system of socio-economic efficiency is also enforced by means of city planning (for example, the use of an orthogonal grid). Thus, architecture becomes part of the civilising mission; a point to which I will return at various points throughout the book.

For Bhabha, the concept of translation is instrumental in developing a deconstructive critique of colonialism …

It is clear that the concept of translation brings forward a wide range of issues regarding the relation between the colonised and the coloniser. Amongst those issues is the hierarchy implicit in the colonial relationship. That is why Homi K. Bhabha finds the term useful in order to challenge the predominance assigned to the 'original', which in his work amounts to the coloniser's culture. For Bhabha, the concept of translation is instrumental in developing a deconstructive critique of colonialism; in other words, a critique which reverses the position of the two primary participants in the relationship: colonised and coloniser. Bhabha's usage of translation as a critical term refers largely to the work of Walter Benjamin, a German philosopher who wrote various essays on literary translation. Benjamin's work, however, has implications on many fields outside literature, including cultural and postcolonial theories, as well as architecture. This chapter elucidates why his work is important and, also, how it is useful in the production of a postcolonial critique.

Walter Benjamin and the 'task of the translator'

The 'Task of the Translator' is one of Walter Benjamin's most celebrated and, indeed, influential essays. It was written as the introduction to his own translation of Baudelaire's *Tableaux Parisiens* into German. Somewhat surprisingly, considering the brevity of the essay, it has become a point of reference in the field of translation studies. As the title indicates, in the essay Benjamin explains how he sees his role as translator. He then proceeds to develop a thesis about the evolution of language and the difficulties it presents for the transmission of meaning from one language to another. The arguments that Benjamin puts forward in his essay, and the fact that it has reached most of its audience through translation (because it was originally written in German), show how translation is an indispensable means for the communication of ideas – as well as for the dissemination of literary works – across languages and cultures. Rather than attempting to summarise its content, I will discuss three aspects of Benjamin's essay that have influenced Bhabha's work: first, that languages are always in a process of historical development (evolution or becoming); second, that languages are linked to one another, a feature Benjamin calls the 'kinship' of languages; third, that translations are independent from the original, or become so once they are produced.

The first of these three aspects is very important in order to understand many issues raised by Bhabha in his writings about colonial discourse. Benjamin argues pointedly that languages are not inert constructions. Quite bluntly, he says that languages are not dead. On the contrary, languages are dynamic systems which change historically and never stop changing. He illustrates this argument by highlighting how expressions that 'sounded fresh once may sound hackneyed later or what was once current may some day sound quaint' (Benjamin 1999: 74). For example, very rarely would someone nowadays shout 'a pox o' your throat, you bawling, blasphemous, incharitable dog' when the bicycle jammed accidentally in the middle of the road. Today, you probably hear a much abbreviated expression: two words, only a handful of letters. The difference between Sebastian's curse to the boatswain in the first act of Shakespeare's *The Tempest* and the most likely insult one would hear nowadays is an example of the way languages change in time, along with cultures. Then, Benjamin argues that considering the dynamism of language it is futile to try and find the essence of language, or any form of idealised meaning in its expressions (words or phrases). To do so is to ignore the fact that languages are always changing. 'More pertinently', Benjamin says, 'it would mean denying, by an impotence of thought, one of the most powerful and fruitful historical processes': the transformation of language (Benjamin uses the expression 'maturing process' of language) (Benjamin 1999: 74). It is important to note how Benjamin refers to the transformation of language as one of the most powerful and fruitful historical processes. In these circumstances, translation becomes very complex because all languages are equally dynamic, not only the language of the original is changing but also that of the translation. So, for Benjamin, translation is more than simply a mechanical act of transmitting information from one language to another. In his own words:

> **translation is as far removed from being the sterile equation of two dead languages that of all literary forms it is the one charged with the special mission of watching over the maturing process of the original language and the birth pangs of its own (Benjamin 1999: 74).**

Consequently, if languages are always changing, then translation can never be total, it is always only provisional; as languages change new translations will be needed. Moreover, Benjamin argues that some aspects of language remain untranslatable and, so, translation is a transitory way of coming to terms with the 'foreignness of languages' (Benjamin 1999: 195). The idea of foreignness refers to the fact that languages are different, and yet interconnected.

The second aspect listed above, namely the kinship of languages, may seem slightly confusing. Although it can be summarised quite simply (as I will do here), the idea of linguistic kinship does open the door onto a great number of theoretical possibilities, some of which I will mention here only for guidance but which would require further analysis if one is to comprehend them fully. For Benjamin, there is a paradoxical relationship between languages in the sense that they seem to exclude one another while, simultaneously, they supplement each other in their intention. By intention Benjamin refers to what languages express or, rather, what people express through language. Benjamin indicates that words used to signify the same object in different languages may have gained alternative sociocultural significance throughout history and, so, when used in literature or in speech, they may intend to signify one thing but the mode of intention may open a gap between the intended meaning and what the listener makes out of them. Benjamin explains his argument by means of a comparison between the words *Brot*, in German, and *pain*, in French – a discussion which Bhabha himself addresses in his essay 'How Newness Enters the World' (Bhabha 1994: 212–35). Although the two words mean 'bread', that is the object they both intend to signify, it could be that the 'mode of intention' (which is determined by a number of social, cultural and historical factors) is different. Consequently, the same word may mean something different to a German than it does to French person – for example, the fact that, in France, bread is most commonly white (i.e. baguette), while in Germany it would most surely be dark (i.e. wholegrain). It follows that the translation of these two words from French into German (or vice versa) cannot be straightforward, because their meaning will be influenced by a number of non-grammatical factors. To put it differently, meaning is not only found in the 'object of intention' (bread), because it is determined by the 'mode of intention' (context). Now, if this is the way in which languages exclude one another – two words that denote the same thing but have acquired different meaning throughout their history – Benjamin argues that they also supplement each other. Supplementarity arises from the fact that 'bread' (the actual loaf) remains the link between the two languages so that any additional meanings these two words have in their respective languages complement each other. To put it simply, by addressing the 'mode of intention' the reader of the translation is made aware of a broader range of social, cultural and historical circumstances which determine the meaning of the word in the original language. It so follows that 'the task of the translator consists of finding the intended effect upon the language into which he/she is translating which produces an echo of the original' (Benjamin 1999: 77). As such, the practice of translation gives agency to the translator to alter the language of original text (its literal meaning, grammar, rhyme, etc.) in order to convey its intended effect in another language. By using the analogy of the echo (or reverberation) Benjamin introduces the notion that original and translation are separate entities which follow different paths, although they are also inseparable.

The third aspect that has influenced greatly the work of Bhabha is the idea that translations become independent from the original. In an evocative passage of his essay, Benjamin maintains that:

> **Just as a tangent touches a circle lightly and at but one point, with this touch rather than with the point setting the law according to which it is to continue on its straight path to infinity, a translation touches the original lightly and only at the infinitely small point of the sense, thereupon pursuing its own course according to the laws of fidelity in the freedom of linguistic flux (Benjamin 1999: 80–1).**

Let us return to Shakespeare's extraordinary curse in the first act of *The Tempest*, 'a pox o' your throat, you bawling, blasphemous, incharitable dog'. One of the reasons why it would not make much sense to proffer such an insult to the distressed rider whose bicycle jammed in the middle of the road is that the pox no longer presents such a threat to his life (moreover, it appears that Shakespeare used the word 'pox' as a generic term for disease). Besides, the religious connotations of blasphemy and unkindness would simply make no sense at all in that context; whereas a loud four-letter word with an arm up in the air would certainly cause a reaction. As stated above, this demonstrates that languages change constantly, but how does it prove that translations ought to pursue their own trajectories? If one tries to translate literally (word by word) either of the two expressions into another language, they would probably make no sense whatsoever and, so, they would not convey an insult. Therefore, as translator, one would need to find an insult in the language of the translation which is commensurate with either of the two insults in English (Shakespeare's or the contemporary slang). In so doing, the translator focuses on the 'mode of intention' rather than the object of intention – pox, blasphemy, unkindness or dog – thereby departing from the original 'in pursuit of its own course according to the laws of fidelity in the freedom of linguistic flux'. In other words, the translation is faithful to the original in trying to convey the intended effect, to insult, but it is also different in its construction (grammar, rhythm, rhyme, etc.) so that the translation is a literary entity of its own. That is why Benjamin affirms that the translation is of no importance to the original – even though they remain connected because the translation would not exist without an original. That is why, in Benjamin's view, the translation is a *re-creation* of the original, not a 'copy'. Benjamin goes even further to assert that 'the life of the originals attains in them [the translations] to its ever-renewed latest and more abundant flowering' (Benjamin 1999: 72). This assertion is of great importance to Bhabha because it dismantles the hierarchy that grants authority to the original over the translation – it is a deconstruction.

When translation is understood as re-creation and reversal it acquires a political dimension suitable to criticise colonial and, even, contemporary systems of domination.

In this case, translations are considered to be the means through which originals continue to live and to transform historically. Thus, translations are a vital and necessary 'part' of the original; vital in the sense that originals continue to live in translation and part (component) because they do need each other in order to exist, or continue to exist. Such a dramatic hierarchical turn could no better suit the work of Bhabha. That is why theories of translation play such an important role in exploring the dynamics of cultural communication in situations of cultural inequality. When translation is understood as re-creation and reversal it acquires a political dimension suitable to criticise colonial and, even, contemporary systems of domination. Understandably, Benjamin's thesis on translation lies at the centre of Bhabha's work.

Translation in Bhabha's work

Let us return to the description of colonialism as a 'translational' phenomenon where everything occurs in and through translation: the translation of people to and from the colony (the physical sense), the subsequent need to communicate in different languages (linguistic translation) and, often, the imposition of educational, religious, economic and political systems which also need to be adapted (cultural translation) in order to operate in different places (colonies or countries) around the world. In these circumstances, translation is not a neutral operation between two equal languages or cultural sites. In situations of colonialism, translation turns into a tool for the construction and exercise of European authority. As in the previous chapter, to develop this idea I will focus on three aspects: first, the representation of the other as inferior; second, the teaching of European languages; and third, the elimination of differences as a means of control. Since I will return to these three issues many times throughout the book, I will not elaborate extensively on them here. Instead, I will only point out how the 'practice' of translation was a tool of colonial domination and how a postcolonial review of the 'theory' of translation can help to develop a critique of such a practice.

… the representation of *primitive* subjects was a way of 'constructing' them both as subjects and as inferior simultaneously.

In order to address the first aspect, let us draw attention to the fact that, since the arrival of the European in the colonies, there was both a need and a desire to provide eager metropolitan audiences with a description of the unknown peoples they had found abroad: their colonial subjects. Such descriptions came in the form of letters, reports, minutes, legal documents, books and so on. To be effective, descriptions need a referent, a familiar image to compare to and, so, to create a clearer picture of the subjects being represented. In this case, the referential point of comparison was the Europeans themselves: their history, arts, fashions, languages, laws and social manners, to mention only a few aspects. Consequently, due to being represented in relation to European norms, colonised subjects could only emerge as the opposite of the European: the naked, black, effeminate, sexualised, uneducated savage. In this sense, the representation of *primitive* subjects was a way of 'constructing' them both as subjects and as inferior simultaneously. Anthropologists in the eighteenth century, who were amongst those trying to generate a body of scientific knowledge about unknown peoples for their audience in Europe – an audience which included the general public as well as scholars – defined this practice as 'cultural translation': the process of translating a culture in terms that are intelligible to members of other cultures. It is clear, however, that this mode of cultural translation produces an asymmetric relationship that grants authority to European cultures. What is more, cultural translation as such justifies colonisation in the sense that it demonstrates the *need* further to impart European knowledge to the backward colonised subject.

Before moving on to the second aspect, it is worth mentioning that this form of cultural representation, or translation, simplifies (or essentialises) colonised subjects and their cultures. The representation of colonised subjects through simplified concepts such as black, effeminate, sexualised, uneducated, insincere or savage is, as Benjamin puts it, a denial of the heterogeneity and historical transformation of languages and cultures. In fact, Bhabha refers to this labelling of the people as 'colonial stereotyping': the fixing of cultural, historical and racial difference to excessively simplified descriptions. For Bhabha (as for Benjamin), the stereotype is not a simplification because it is untrue, it may well be an accurate representation of an object/subject (for example, some colonial subjects are actually black). The problem arises because the stereotype simplifies the represented subject reducing it to a self-contained, complete, immutable assemblage and, so, it ignores the historical processes through which such an assemblage has been formed. To put it simply, the problem of colonial stereotyping is not that it represents subjects as, say, black, but the fact that it implies that all Blacks are the same. In that sense, colonial cultural translation dissociates its subjects from their own history in order to introduce them in the European history as recognisable objects whose difference ratifies the superiority of the coloniser and, thereby, validates colonial intervention.

This takes us to the second aspect: education and the teaching of European languages. Since it was established a priori (without need for justification) that colonised subjects were inferior, colonial intervention was necessary in order to demonstrate to them the benefits of western civilisation. That is why education in the European languages and the European way was imparted in most colonies. It is worth mentioning that the responsibility for education rested largely in the hands of missionaries (especially in English, French, Portuguese and Spanish colonies) and, so, European education served to accomplish many goals simultaneously: to pass on European knowledge, to make indigenes appreciate the greatness of western civilisation and to convert colonised subjects to Christianity. As Macaulay put it in his infamous Minute of 1835, which I will discuss at length in Chapter 4:

> **the function of English education is to form a class who may be interpreters between us (the British) and the millions whom we govern; a class of persons Indian in blood and colour, but English in taste, in morals and in intellect (Macaulay 1835).**

One of the most important tasks undertaken by missionaries in the colonies was the translation of the history of colonised subjects into European languages. In this context translation refers both to linguistic translation (from indigenous languages to European) and to the actual writing of history; because in certain parts of Africa and the Americas, histories were not written, they were oral. Thus, missionaries produced textualised versions of the history, technology and religion of colonised subjects, which were written according to the western classical scholarly canon (that is, chronologically linearly) and, then, read back that history to the colonised as part of their education. In Colombia, for example, Catholic priests taught the Muisca about Bochica in Spanish – Bochica was the deity who, according to Muisca tradition, drained the valley in the Andes where they resided, the site of the current capital Bogotá. At the same time, the Muisca were taught about the scientific improbability of such a story. In this case, colonised subjects got to 'know' about themselves through translation – in the language and through the educational methods of the coloniser – and received instruction as to their position in relation to the more advanced culture of the coloniser. The opposite operation also took place. In many cases, missionaries learned the languages of the colonised and wrote 'grammars' and 'dictionaries' to be able to speak and to teach in those languages. In the process, missionaries rationalised the languages of the other and incorporated them into the European system. These two forms of translation served the purpose of disseminating the superior culture of the European.

<u>To impose metropolitan languages, knowledge, taste, morals and so on, is to transform colonised subjects into copies of Europeans.</u>

Thus we arrive at the third aspect listed above, the elimination of differences as a means of control. From the previous two aspects we can already identify two forms of homogenisation. First, we referred to the practice of stereotyping, or the representation of colonised subjects in terms of over-simplified images which show them as homogeneous assemblages; an undifferentiated mass of black, effeminate, sexualised, uneducated, insincere people (to mention only a few appellatives). This kind of cultural representation disregards the history of the subjects being represented and, hence, erases all the differences that exist inside their sociocultural organisation. Second, we examined another form of homogenisation, achieved largely through education. To impose metropolitan languages, knowledge, taste, morals and so on, is to transform colonised subjects into copies of Europeans (or the attempt was made to do so). While in the first case the homogenisation of colonised subjects helps to make them intelligible to a European audience, the second serves various other purposes: facilitating governance, optimising productivity and stimulating trade. It so transpires that the practice of colonial translation was as much a means to facilitate cultural communication as it was a strategy of containment. In other words, in the colonial context, translation works both as a way to construct authority and to exercise it. As such, colonial translation is different from Benjamin's deconstructive understanding. Rather than serving as a vehicle to construct and exercise authority, Benjamin proposes that translation is a practice that serves to unsettle the predominant position of the original in relation to the translation; or, indeed, the predominance of European culture over the cultures of colonised subjects. And so, this discussion should also help to understand why Bhabha turns to Benjamin in order to develop a critique of colonial translation.

Bhabha has not produced a comprehensive study of translation, nor has he developed a postcolonial theory of translation – which could be found, rather, in Tejaswini Niranjana's compelling book *Siting Translation: History, Post-Structuralism and the Colonial Context* (1992). However, Bhabha resorts to Benjamin's work on translation in order to bring to light those areas of contradiction (or ambivalence, a term explained in the next chapter) inherent in the practice of colonial translation described above. Bhabha is particularly interested in the notion of supplementarity where languages do complement one another through their differences rather than their similarities. That is because the 'mode of intention' opens the door onto a whole set of historical and cultural circumstances that go beyond the mere meaning of the word. At the same time, this reveals that there are elements in language that cannot be translated, they are untranslatable and, so, languages remain foreign (like *Brot* and *pain*). Bhabha capitalises from this contradictory characteristic of languages in order to propose that Benjamin's argument can be used in order to develop a theory of cultural difference (see Chapter 6). Rather than striving towards the elimination of difference, as explained above, Bhabha suggests that in the process of translation, content (whatever is translated, a language, a history, a culture) becomes alien and estranged. By way of example we can return to the story of Bochica, a story that was translated into Spanish by Catholic missionaries and then repeated back to the Muisca, for whom Bochica was a deity. That is precisely the moment when content becomes alien and estranged. It is not that the story itself changes but its significance and the position it occupies within the Muisca system of beliefs does change, becomes estranged. Consequently, Bhabha says that the language of translation is confronted with the untranslatable which, in this case, is the significance of that story in the Muisca mythology – an aspect that cannot be represented in language or translated into Spanish and, hence, cannot be fully integrated in the rational structures of the coloniser. Translation remains incomplete and, hence, rather than erasing, it exacerbates difference; it produces a third, or hybrid, story that

> ... Bhabha affirms that 'translation is the performative nature of cultural communication'.

is neither the original Muisca myth nor a western piece of literature. In fact, untranslatability reveals the unbridgeable gap between the two cultures. This gap between cultures, however, is not a void; it is the space where cultural meaning is constantly negotiated and reconstructed. That is why Bhabha affirms that 'translation is the performative nature of cultural communication' (Bhabha 1994: 228). In so doing, Bhabha locates the production of cultural meaning in the realm of the untranslatable, the interstices between languages and cultures – not in the centre, or, in this case, in the hand of the missionaries.

> The political dimension of the notion of translation in Bhabha's writings is more clearly perceived in his discussion about migrant minorities living in the west today.

The political dimension of the notion of translation in Bhabha's writings is more clearly perceived in his discussion about migrant minorities living in the west today. He describes the migrant's experience as caught between national affiliations and metropolitan assimilation policies, a situation for which there is no actual resolution. In practical terms, irresolution is seen in the fact that the migrant is caught between his or her 'past' (country of origin and nationality, language, customs, and so on) and his or her 'present' as a resident of another country with different culture and traditions. Even if a migrant obtains nationality in his/her country of residence, and so become legally British, Canadian or French for example, he or she remains different, or differentiated, by his or her accent, ethnicity, manners and social or familial affiliations. Rather than being assimilated or naturalised according to legal terminology, the migrant remains somewhere in the middle between cultural sites. Thus, the migrant emerges as the subject of cultural difference, the element in translation that does not lend itself to translation but remains in a situation of in-betweeness. In Bhabha's own words:

> The migrant culture of the 'in-between', the minority position, dramatises the activity of culture's untranslatability; and in so doing, it moves the question of culture's appropriation beyond the assimilationist's dream, or the racist's nightmare, of a full transmittal of subject matter and towards an encounter with the ambivalent process of splitting and hybridity that marks the identification with culture's difference (Bhabha 1994: 224).

As is clear, in Bhabha's view, migrant minorities do not lend themselves to cultural translation in the way one would expect languages to be translated – as a full transmittal of subject matter. Instead, migrant minorities remain in a state of in-betweenness, as 'stubborn chunks' which never blend with the host culture but can never be reconstituted as they were prior to migration. Therefore, they seem not to belong to any particular cultural system but partake in all of them as hybrid cultural elements that are both different and differential. As we will see later, hybridity can be considered the result of cultural translation or, indeed, evidence of its impossibility. Hybridity highlights the foreignness of cultural languages and, at the same time, demonstrates the dynamism of translation as the staging of cultural difference.

Hybridity highlights the foreignness of cultural languages and, at the same time, demonstrates the dynamism of translation as the staging of cultural difference.

I expect to have shown two key dimensions of the discussion about translation. On the one hand, colonial translation as a mode of representation whose objective is to construct colonised subjects in a manner that is both intelligible for the European and inferior in relation to them. Inferiority justifies colonisation and the perpetuation of European authority. On the other hand, there is postcolonial translation, which theorises the concept of translation in order to advance a critique to its colonial practice. I have focused only on the work of Benjamin, whose ideas are recurrent in Bhabha's writing. However, other critics have also worked extensively on the concept of translation and on the writings of Benjamin, for example, Jacques Derrida and Paul de Man. To be sure, Derrida and de Man have also influenced Bhabha's usage of translation and his postcolonial critique. In spite of their importance in the context of translation studies, I have not reviewed their work because it would have implied too great a theoretical deviation; mostly away from postcolonial theory. Notwithstanding their exclusion, I feel compelled to do justice to their work by pointing again in the direction of Tejaswini Niranjana who has carried out an exhaustive study of Benjamin, Derrida and de Man as the three figures who have helped to shape contemporary debates about cultural and literary translation. In fact, Niranjana argues that:

> The rethinking of translation [from a postcolonial perspective] becomes an important task in a context where it has been used since the European Enlightenment to underwrite practices of subjectification, especially for colonised peoples. Such a rethinking – a task of great urgency for a postcolonial theory attempting to make sense of 'subjects' already living 'in translation', imaged and re-imaged by colonial ways of seeing – seeks to reclaim the notion of translation by deconstructing it and re-inscribing its potential as a strategy of resistance (Niranjana 1992: 6).

For Niranjana, resistance is a deliberate and interventionist – deconstructive – act of translating history that is no longer concerned with the universalising and homogenising agenda of western cultural politics, but with the acknowledgement of difference and cultural heterogeneity. For Niranjana, to 'translate history' implies carrying out a radical rewriting of it from the perspective of previously colonised peoples. Indeed, the task that Niranjana urges be undertaken is of great relevance in architecture, a field whose history has been written almost exclusively from the perspective of Euro-American academia – as will be shown in the next chapter.

CHAPTER 3

Ambivalence

The previous chapter explores the implications inherent in the colonial 'practice' of linguistic and cultural translation. It highlights the need to re-theorise the 'concept' of translation from a postcolonial perspective in order to overturn the derogatory effects of its colonial usage. Additionally, the previous chapter offers a general overview of the underlying principles of colonial discourse. By colonial discourse I refer to the entire body of knowledge and technology, as well as the methods of cultural representation used to construct and dominate colonial subjects. In general, the previous discussion can be taken as an extension of the introduction leading into more specific areas in the work of Bhabha. This chapter explains another concept which underpins Bhabha's critique of colonial discourse: ambivalence. Although he is more readily associated with terms such as hybridity, mimicry and the performative, the critical capacity that he assigns to these terms rests largely on the psychoanalytic concept of ambivalence. By explaining the concept of ambivalence, this chapter continues to build up a basic theoretical context before undertaking the analysis of the previously mentioned three concepts: hybridity, the Third Space and the performative.

Psychoanalysis, as a method of studying and understanding the mind, has had an immense impact in fields other than psychiatry. It has influenced the work of countless twentieth-century thinkers, particularly those known as 'post-structuralists'. According to psychoanalysis, the mind is not a complete and coherent assemblage which develops linearly in time as people grow older.

Instead, psychoanalysis proposes, and demonstrates through the study of clinical cases, that the human mind is fragmented at all stages during its formation (from birth to adulthood) and its development is non-linear – the mind contains, and reverts back to, past experiences either consciously or unconsciously. In turn, these disorderly (non-linear) experiences determine our behaviour. Thus, according to psychoanalysis, our identities are not single, nor are they static. On the contrary, identities are plural and always changing, even our own individual identities display multiple facets. That is why psychoanalysis provides a suitable theoretical setting to question many assumptions about the notion of identity, both individual and collective, according to which identities are homogeneous and develop linearity following the pattern of classical western historicity; two issues that are acutely challenged by postcolonial critics.

What emerges, instead, is a subject caught in-between, simultaneously integrated and rejected by the dominant system ...

The use of psychoanalytic theories is not limited to the work of Bhabha. Frantz Fanon, Stuart Hall, Edward Said and Gayatri Spivak, for example, also resort to psychoanalysis in order to advance critiques of colonialism, as well as other forms of domination, such as neocolonialism and imperialism. Amongst these critics, the work of Fanon is particularly interesting because he was a practising psychoanalyst. Considering that the main theme of this chapter is the notion of ambivalence, it is worth mentioning one of Fanon's various books, *Black Skins, White Masks*, a book whose title encapsulates the sense of colonial ambivalence postcolonial theorists, like Bhabha, refer to in their writings. The title of Fanon's book presents a contradictory condition in the position of the black self in relation to the white other. Put simply, it reveals an internal conflict in the colonised subject who speaks, behaves, dresses like the white 'master' (that is the mask) but remains differentiated (discriminated against) by the darker colour of her or his skin. So, being black remains a sign of difference and a reminder of inferiority. In that sense, the title of Fanon's book (and indeed the book in general) proves the impossibility of fully realising the colonial enterprise of creating subjects whose skin colour is different but who are European in their education, manners and taste. What emerges, instead, is a subject caught in-between, simultaneously integrated and rejected by the dominant system and unable to return to any previous state prior to colonisation; a state from which colonial subjects have been removed permanently. It is precisely this contradiction in the discourse of colonialism that constitutes Bhabha's ambivalence.

It is precisely this contradiction in the discourse of colonialism that constitutes Bhabha's ambivalence.

It is important to note that Fanon focuses on the internal conflict experienced by the colonised subject, the desire to be white and the acknowledgement of difference which prevents the realisation of that desire. Bhabha, on the other hand, situates ambivalence in the discourse of colonialism, that is, in the contradiction between the colonisers' desire to see itself repeated in the colonised and the rejection of that repeated other – the translation, or the copy – in order to keep their authority. In Bhabha's view, this ambivalence shows that the colonisers are also internally in conflict between their wish to repeat themselves in the colonised (what Bhabha calls the 'narcissistic demand') and the anxiety of their disappearance as a result of the repetition; because if the Other turns into the same, difference is eliminated, as are the grounds to claim superiority over it.

As mentioned above, psychoanalysis argues that no identity is single and complete but, on the contrary, that all identities are torn within themselves, inhabited by internal conflicts and contradictions: they are ambivalent. This explains one of Bhabha's most celebrated contributions to postcolonial discourse: the dismantling of the straightforward dichotomy between self and other, represented in the identities of the colonised and the coloniser. That is also the reason why the concept of ambivalence underpins Bhabha's discussions on hybridity, mimicry and the Third Space (discussed in Chapters 4 and 5), as well as his description of the split temporality of the nation – which he calls the pedagogical and performative (discussed in Chapter 6). Departing from the premise that authoritarian discourse requires univocality – a single unambiguous voice – in order to be authoritative, by revealing the ambivalence inherent in both the discourse and the practice of colonialism, Bhabha theoretically undermines the validity of the coloniser's claim for uncontested authority.

Postcolonial ambivalence

Let us explore the concept of ambivalence in psychoanalysis before returning to the way in which Bhabha uses it in his writings. In psychoanalysis, ambivalence refers to the coexistence of two contradictory instincts, or desires, particularly love and hate. Ambivalence is different from indecision, wanting one thing while simultaneously wanting another (for example, difficulty deciding whether one wants coffee or tea, jeans or chinos, straight or curved walls), which in psychoanalysis is referred to as an 'ambivalent oscillation' rather than ambivalence. An early manifestation of ambivalence occurs during the Oedipus phase: when the child adores one of his parents and, consequently, dislikes the other – boys adore their mothers and see their fathers as sexual rivals while girls love their fathers and perceive their mothers as contenders for their love object. Of course, the child does not only hate her or his sexual rival, the child also loves the rival at the same time; the child recognises that her or his rival is, after all, a parent. That is why Freud argues that the Oedipus complex is a compelling example of ambivalence because it shows an irresolvable coexistence of love and hate.

Freud also uses the notion of ambivalence in his book *Civilisation and Its Discontents* (2002) in order to discuss how people in certain conditions of adjoining territoriality relinquish the satisfaction of their tendency to aggression in order to attain a measure of security. Taking the Spanish and the Portuguese, as well as the English and the Scottish, as examples of an ambivalent relationship of love and hate, Freud argues that 'it is always possible to bind quite large numbers of people together in love, provided that others are left out as targets for their aggression' (Freud 2002: 50). As in the rest of the book, Freud is supporting his case that civilisation is built on repression. Civilisation limits people's desires and represses their tendencies to aggression, sexual gratification and other forms of obtaining pleasure. In other words, to achieve civil coexistence it is necessary to repress some of mankind's natural tendencies, drives and desires. Therefore, there is an inherent ambivalence in the principles of civilisation because it requires the repression of the individual in order to achieve collective security (welfare), but those repressions also cause discontent amongst individuals. At this point Freud underlines a complication: repressed tendencies do not simply disappear, they are diverted, channelled in other directions so that they can be released. In this case, Freud suggests that the tendency to aggression is released on other social groups outside the relationship between two neighbouring peoples; others outside become the objects of aggression.

It may now be clearer why Bhabha brings up this passage from Freud in his writing about colonial ambivalence. The argument can be developed that in order to guarantee a certain degree of civil and peaceful coexistence in Europe, non-European peoples became the targets of European aggression – and, indeed, continue to be today. However, history shows not only that there were significant rivalries between countries in Europe (as there are now), but also that the countries themselves were/are not homogeneous and stable socio-political organisations. Thus, it arises that the claim for cultural authority was never 'univocal' (single-voiced and unambiguous) but ambivalent – that is, multi-vocal, internally broken and in conflict.

Let us now examine the role of the psychoanalytic notion of ambivalence in Bhabha's critique of colonial discourse. For Bhabha, colonial discourse is characterised by an inherent contradiction, an ambivalence which occurs in the process of constructing authority through the representation of colonised subjects. As explained above, colonial authority is constituted primarily in two ways. On the one hand, it exists a priori, either it needs no justification or its evidence comes from within itself (i.e. European architecture is superior to the Aztec because Aztec buildings are analysed using European methods of analysis). On the other hand, authority is built through strategies of discrimination, that is, by establishing a difference between two parties

This form of ambivalence, which lies at the centre of Bhabha's discussion on hybridity and mimicry, can also be found in the construction of other authoritarian discourses like architecture ...

in such a way that one emerges as superior (i.e. European people are superior because they are civilised, white, masculine and dress in a certain way, while the colonised are inferior because they are uncivilised, darker, effeminate and do not wear clothes). Clearly, these two modes of constructing authority are not separate but interconnected. Since authority is constituted a priori, the colonisers take it as their duty to 'civilise' their subjects, make them into a double image of themselves. Yet, when colonised subjects attain a certain similarity – when they speak European languages, wear clothes and live in a 'civilised' manner – they need to be discriminated against in order for the coloniser to maintain their authority. They do so by stressing differences like ethnicity, accent or the dress codes of colonised subjects, as well as the form of their buildings. Then, colonised subjects, who were expected to become 'like' the European, are simultaneously rejected, differentiated or disavowed. This simultaneous operation of identification and disavowal is the basic form of ambivalence to which Bhabha refers in his writings. In his view, this contradiction disrupts the authority of colonial discourse because it proves that the claim for authority is equivocal, it shows the coexistence of ambivalent and contradictory desires (the wish to see themselves repeated in the colonised and the need to repudiate that image). This form of ambivalence, which lies at the centre of Bhabha's discussion on hybridity and mimicry, can also be found in the construction of other authoritarian discourses like architecture, as will be demonstrated in the final part of this chapter.

So far we have discussed only a basic form of ambivalence, namely the straightforward coexistence of two opposite instincts or desires, like in the

Oedipal phase mentioned above. However, Bhabha also addresses another form of ambivalence, namely the conflict between the principles of civilisation and the discontent those principles cause amongst 'civilised' individuals. Unlike Freud, though, Bhabha is not interested in the concept of civilisation as such, but in the figure of the nation, which is considered one of the greatest achievements of modern civilisation. Bhabha thus asks whether 'the people are the articulation of a doubling of the national address, an ambivalent movement between the discourses of pedagogy and performative?' (Bhabha 1994: 149). With this question, Bhabha points out that the nation, as an instance of civilisation, is a pedagogical construction which represents people as a homogeneous body, an undifferentiated mass. That is why, for Bhabha, the nation is taken as a mechanism that erases cultural differences for the sake of maintaining social order, economic prosperity and governance. As a rational construction the nation ignores the complex realities of the people, their convoluted histories and their multiple modes of socio-political expression. Bhabha maintains that in the process of building the nation, people in general (not only colonised subjects) are removed from their multiple cultural positions (i.e. ethnicity, gender and class) and re-inscribed in the homogeneous national community. Consequently, people are the articulation of the desired image of a homogeneous national community and the heterogeneous reality of the multicultural modern nation in a globalised era. In Bhabha's view, this form of ambivalence renders questionable the narrative address of the nation (the discourses that conceive it as homogeneous, stable and sovereign). Instead, he presents the nation as an unrealisable project always in the process of being made (see Chapter 6).

Another figure who has influenced Bhabha's use of psychoanalysis is Jacques Lacan. In fact, some commentators argue that Bhabha makes more extensive use of Lacanian psychoanalysis than he does of Freudian. The argument may well be true considering the great deal of influence that Lacan has had on literary studies since the 1970s, as well as on cultural theory. I will not elaborate extensively on Lacan in this chapter because a further explanation is given in Chapter 5. However, a brief summary of the reasons why Lacan's theories have influenced literary criticism may be useful to understand other aspects of

Bhabha's work. Let us start this brief summary by looking at language. Language occupies a privileged position in Lacan's theory on the formation of identity. Unlike Freud, whose theory on identity relies heavily on biological differences (i.e. male and female identities are determined by the presence or absence of the phallus), for Lacan, language is also an important determinant of identity because language precedes the self; the system of language was there before one was born. Consequently, language contributes to the formation of identity because we learn who, or what, we are by the way people refer to us through language. For example, the use of pronouns like he or she, which specify gender, play a central role in the formation of our identities; we learn through language that we are man or woman so that gender differences are as external to us as they are biological. Though this is a simplification of Lacan's theory, which is more complicated than the use of pronouns, I hope it helps to illustrate the fact that Lacan assigns a great deal of importance to language. When taken to the realm of literature, Lacan's theory makes it possible to shift the focus of literary criticism. Instead of trying to identify, or interpret, the unconscious of the characters and the author of the book, attention is given to the text itself and the relationship between the text and the reader. As explained in the Introduction, this forms the basis of deconstruction, the reversal of traditional hierarchical structures that give predominance to the author (and the characters in the book) over the reader, who is expected to understand the author's intention. In other words:

> the object of [literary] psychoanalytic criticism was no longer to hunt for phallic symbols or to explain Hamlet's hesitation to revenge his father's death by his repressed sexual desire for his mother, but to analyse the way [in which] unconscious desires manifest themselves in the text through language (Homer 2005: 2).

This is, largely, what Bhabha does in all his analyses of colonial texts: he reveals the intricate play of unconscious desires, traumas and ambivalences that are expressed in the language of the text. Bhabha, however, does not focus exclusively on texts of literature (say, fiction) but in all forms of literary expression. He employs methods of literary analysis on biographical, historical and legal texts, amongst other genres, in order to reveal their intentions and implications in the construction and exercise of colonial authority. That is why Bhabha's reading of European texts is often referred to as a 'deconstructive reading': because it unsettles the predominance given to the author and reassigns it to the postcolonial reader or critic (the notion of translation and the task of the translator examined in the previous chapter are useful to understand this point). Also, by employing terms such as ambivalence (and splitting, doubling and partiality, which were developed in psychoanalysis by Lacan), Bhabha dismantles the bipolarity of literary and cultural analysis (as explained in the Introduction).

<u>That is why Bhabha's reading of European texts is often referred to as a 'deconstructive reading': because it unsettles the predominance given to the author and reassigns it to the postcolonial reader or critic.</u>

Three aspects have been exposed in relation to Bhabha's use of psychoanalytic ambivalence: first, it reveals the contradiction inherent in the construction and exercise of colonial authority; second, it helps to challenge other narratives of power, like nationalism; and, third, it assists in carrying out deconstructive readings of European narratives and texts (i.e. history). Even though these three aspects disrupt the construction and exercise of authority, the theoretical reversal they help to carry out appears to be distant from the actual focus of criticism: people (colonised subjects, minorities, etc.). Can this reversal take place politically, that is, can colonised subjects be relocated in a position of authority? According to Bhabha it is possible. He argues that 'the ambivalence at the source of traditional discourses on authority enables a form of subversion, founded on the undecidability that turns the discursive conditions of dominance into the grounds of intervention' (Bhabha 1994: 112). In other words, not only does ambivalence disrupt the authority of colonial discourse, it also opens up a space of contestation and resistance which makes authoritarian discourse vulnerable. This is a point of paramount importance because it assigns agency to people to resist and react against the imposition of power. However, Bhabha has been criticised because he does not specify how exactly colonised subjects detect the ambivalence he refers to and turn it to their advantage. Robert Young, for example, has asked whether the outcome of Bhabha's psychoanalytical critique is a theoretical construct that can only be produced in the context of contemporary academia. He says:

> While on the one hand Bhabha suggests that the ambivalence of colonial discourse allows exploitation so that its authority can be undermined even further, on the other hand it seems also to be implied here that such a slippage does not so much occur under the historical conditions of colonialism, but is rather produced by the critic (Young 1990: 155).

Although Young acknowledges that 'a psychoanalytic reading enables contestation of the claims of any [cultural] Europeanisation' (Young 1990: 155), his incisive critique is legitimate, especially considering Bhabha's lack of historical analysis (a point which, indeed, has been criticised by others, as will be noted in the next chapter). In Young's view, Bhabha's highly theoretical analysis is so far removed from the realities of specific cultural groups that it is difficult to imagine how it could be realised in practical terms, not only theoretically. To turn effectively the conditions of dominance into grounds of intervention, as Bhabha suggests, it is necessary to focus on specific locations both historically and geographically. By way of example, I will refer to a case of admitted missionary failure in eighteenth-century Nueva Granada, the name given by the Spanish to their colonies in the north of South America. In his essay 'Guajiro Culture and Capuchin Evangelisation: Missionary Failure on the Riohacha Frontier' (1995), Lance Grahn demonstrates through extensive historical evidence that the Guajiro Indians regarded the missionaries with great distrust due to the 'overt irony of the Christian teaching of love versus Christian brutish behaviour' (Grahn 1995: 132). It was reported by Spanish missionaries at the time that indigenous leaders saw this contradiction as a reason to reject both the teachings of the European missionaries and the conditions of governance they were trying to impose. Indigenes also questioned the fact that they had been taught to farm and trade livestock but were not permitted to benefit economically from it in the same way the coloniser did. The Guajiro Indians, then, turned the discursive conditions of dominance into grounds of intervention. As Grahn points out:

> Indian reactions to European intrusions also illustrated the capacity of Native Americans to use imperialism against itself accommodating some elements to combat what they deemed its more destructive components. Guajiro headmen quickly learned the techniques of European trade and realised its benefits. They not only mythicised European-introduced livestock but also used it in profitable commercial intercourse with British and Dutch traders, establishing a useful and powerful counterweight, including access to superior weaponry to Spanish influence in the area. This same trade supported regional contraband trafficking, of which Spanish colonists and administrators took advantage. Guajiro participation in Caribbean trade networks, then, enabled clans to resist Spanish subjugation (Grahn 1995: 130–1).

Grahn's account of Guajiro resistance can be seen as an example of what Bhabha calls cultural hybridisation (see Chapter 4); a process which represents a powerful site of cultural productivity that destabilises both the discourse and the exercise of authority. Grahn proves that Guajiro Indians detected the ambivalence inherent in the colonial discourse and, hence, resisted it – they appropriated some traits of the coloniser and used them to their advantage. This example also responds to Young's enquiry as to whether ambivalence was detected by colonised subjects during the colonisation or whether its effect is produced by the critic in the present. Young's criticism remains valid in the sense that, to be effective, certain forms of theoretical criticism require historical contextualisation – and it is missing in Bhabha's work.

However, this is not the only aspect of Bhabha's use of psychoanalysis which has raised concerns amongst critics. An issue one encounters often when reading critiques of Bhabha's work is that he is unorthodox in his appropriation of psychoanalytic theory. Scholars of psychoanalysis accuse Bhabha of being

Indeed, Bhabha's usage of psychoanalytic ambivalence in postcolonial theory appears to be useful in producing a critique of architecture.

extremely teleological, or purpose driven, in reducing a problem that is broad and complex, with more to it than the coexistence of opposite instincts within the same psychic space, in order to suit his postcolonial theory. Concerns have also been raised about the way Bhabha moves swiftly between Freud and Lacan without sufficient differentiation. In fact, practising psychoanalysts complain of the lack of rigorous scientific method in Bhabha's use of psychoanalysis. I mention these critiques not in order to undermine Bhabha's work, but in order to make the reader aware of the various aspects that have been challenged and to encourage further understanding of Bhabha through the study of other texts. Despite critiques about Bhabha's appropriation of psychoanalysis, the way in which he uses the concept of ambivalence has provided an opportunity to develop a critical assessment of authoritarian discourse, not only colonial but also other narratives of authority, like nationalist discourses. Indeed, Bhabha's usage of psychoanalytic ambivalence in postcolonial theory appears to be useful in producing a critique of architecture.

The ambivalence of architectural history

At this point I would like to turn to architecture, for it too belongs to that category that Bhabha calls 'discourses on authority'. As such, architectural discourse also presents an ambivalence that is more strongly detected in the writing of history. That is why I propose to examine the way in which non-western architectures have been inscribed in the distinctly singular 'history of architecture'. The following discussion can be seen as a 'postcolonial reading' of history following the procedures explained so far. The purpose of this reading is to disclose the intentions, implications and contradictions inherent in the historical text; a text which serves to grant authority to architectural discourses and buildings produced in the west. Put simply, a postcolonial reading of architectural history would place under scrutiny the authority of architectural history as another kind of Europeanising discourse; a discourse which, like colonial discourse, is characterised by a simultaneous operation of inclusion and exclusion. It may not be possible to undertake such a gigantic project within the confines of this brief explication on the work of Homi K. Bhabha. However, it is possible to identify the reasons why such a task is not only necessary but urgent.

The writing of history, which manifests itself in the architectural history book, becomes the means to create and maintain the referential system with which to judge architectural production around the world, linking with Europe's past. The book records only those case studies that meet the parameters of the system and, of course, excludes buildings that do not comply with it. Therefore, it is no surprise that non-western architectures are customarily represented in the history book through buildings that are similar to European or North American referents. Not only is this a way of establishing a lineage or genealogy, it is also a strategy that gives authority to European architecture (either classic or modern). In general, non-western architectures are celebrated by the historian only when they reach a high degree of refinement in relation to the European architectural canon. It is not the buildings' intrinsic qualities that matter to the historian, or the way buildings respond to the particular needs of users in specific parts of the world, but the fact that they comply with hegemonic architectural narratives – indeed, the history book constitutes a hegemonic architectural narrative. In this manner, the architectural history book tells us much more than simple illuminating facts about buildings around the world.

<u>In general, non-western architectures are celebrated by the historian only when they reach a high degree of refinement in relation to the European architectural canon.</u>

The book itself becomes a piece of evidence that incriminates the historian in the construction of hegemonic architectural narratives. The book reveals that architectural history is constructed through a systematic process of elimination of difference which allows the authorisation of a single architecture to be representative, to become dominant. As a pedagogical construction, or dominant narrative, architectural history is only concerned with the disciplinary agendas of architects and historians, not with the interests of the people in general (I will return to this argument repeatedly throughout the book). As a result, non-western architectural practices are inscribed in the history of architecture but they always emerge as both posterior and inferior in relation to the European or North American predecessor. More dramatic, however, is the fate of those architectural practices which are excluded from the history book altogether, whose exclusion amounts to academic inexistence.

The discriminatory rhetoric, and choice of words used to refer to non-western buildings, is indicative of the position occupied by the historian, as well as the position assigned to European and North American architectures, in relation to the non-western. In her book *Imperial Eyes: Travel Writing and Transculturation*, for example, Mary Louise Pratt offers an interesting reading of Alexander Humboldt's notes on the architectures he found while travelling through the Americas. Pratt shows how Central American indigenous architecture is rendered imperfect, as well as aesthetically incompetent, in relation to the western classical canon:

> **American architecture, we cannot too often repeat, can cause no astonishment, either by the magnitude of its works or the elegance of its form, [Humboldt] writes, but it is highly interesting, as it throws light on the history of the primitive civilizations and the inhabitants of the mountains of the new continent. While in Greece, religions became the chief support of fine arts; among the Aztecs, the primitive cult of death results in monuments whose only goal is to produce terror and dismay (Pratt 1992: 134).**

Pratt's main point is that 'the European imagination produces archaeological subjects by splitting contemporary non-European peoples off from their pre-colonial, and even their colonial pasts' (Pratt 1992: 134). Such splitting occurs during the construction of a history of subjects who did not have a history and who only attained historical subjecthood through the scholarly historicising methods of the coloniser. In other words, colonised people are withdrawn from their own history in order to be inserted into the western linear history devised to grant itself authority. The ambiguity lies in the fact that Humboldt recognises the existence of an indigenous past – he says that indigenous architectures throw light on the history of primitive civilisations – yet that past is subsequently dismissed because it does not correspond to the European classical canon and, also, because it is presumably unknown to Humboldt. Another form of splitting is found in the simultaneous recognition and disavowal of architectural difference. In his notes, Humboldt affirms that indigenous American architectures are not comparable, in scale and image, with Greek and Roman classic architectures, which are taken as superior referents. Thus, indigenous architectures are recognised by the historian, Humboldt in this case, but are immediately declared inferior. The incoherence of these two forms of splitting, the simultaneous inscription and rejection of pre-colonial buildings in a homogenising universal history, unveils the ambivalence of colonial discourse in the historicisation of architecture.

... colonised people are withdrawn from their own history in order to be inserted into the western linear history devised to grant itself authority.

The same operation of identification and disavowal carried out by Humboldt in the context of Central and South American colonialism, and in relation to pre-colonial architectures, has been repeated recently in relation to modern architecture in non-western countries. This time, the classical canon is replaced by modern architecture, which becomes the new referent. As such, modern architecture resorts to a linearity of architectural production whose origins can be traced to European industrialisation and whose genealogy includes the work of a select group of architects: Le Corbusier, Mies van der Rohe,

Frank Lloyd Wright, to mention only some. Despite the 100 years that separate us from Humboldt's derogatory description of indigenous American architectures, contemporary architectural historians continue to use the same rhetoric in order to belittle architectural production in the so-called developing world.

In order to expand on this point, I will focus on William J. R. Curtis' book *Modern Architecture since 1900*, which looks at the chronological 'evolution' of modern architectural ideas in Europe and examines the way such ideas travelled to other locations around the world. Due to the linear historical structure of the book, there is no significant discussion of non-western architectures until Chapter 21, where Curtis talks briefly about the spread of modernist ideas to northern Africa, the Middle East, South Africa, Brazil and Mexico (please note that I refer here to the expanded edition reprinted in 2000). Only in Chapter 27 does Curtis refer to non-western architectures in greater detail. In an earlier edition, the chapter was entitled 'The Problem of Regional Identity'. However, it was changed in later editions to 'The Process of Absorption: Latin America, Australia, Japan'. Despite the title change, and the apparent geographical focus of the newer title, the introduction to the chapter remained unchanged. In the first paragraph Curtis specifies that the modern movement in architecture was 'the intellectual property of certain countries in Western Europe, of the United States and of some parts of the Soviet Union' (Curtis 2000: 491). It is not necessary to subscribe to any particular theory to find this statement provocative, yet, in the context of contemporary cultural studies and postcolonial theory it is very problematic. With the use of legal terminology Curtis assigns the indisputable right of authorship to a select group of countries which become the only possessors of modern architecture.

Subsequently, Curtis proceeds to explain that, 'by the end of the 1950s, *transformations*, *deviations* and *devaluations* of modern architecture had found their way to many other areas of the world' (Curtis 2000: 491; my emphasis). The expression 'found their way to many other areas of the world' is particularly interesting. On the one hand, it implies difficulty in the sense that it stresses the length and arduousness of a journey to other parts of the world. On the other hand, it suggests that the transformations, deviations and devaluations found their way to other countries by themselves. However, throughout the entire chapter Curtis is at pains to stress the fact that developing countries *received* modern architecture from Europe mainly via the work of Le Corbusier. So, it is clear that for Curtis the dissemination of modern architecture follows a genealogy which finds its roots in Europe and develops via the work of an exclusive selection of European and North American architects whose work becomes the referent against which non-western architectural production is judged. Consequently, this select group of western architects is put forward as the agents of dissemination, the medium through which modern architecture 'found its way' to the developing world – those architects represent architecture's 'civilising mission', the bearers of progress via functionalism and the aesthetics of purism. Rather than referring to the arduous geographical journey from the centre to the periphery, the expression 'found their way' reveals an anxiety at the moment when modern architecture escaped the exclusive control of western architects – and, so, threatened the authority of Curtis' select group of architects and the countries who own the intellectual rights of architectural modernism. In other words, once the non-western had appropriated the terms of western architecture, its discourse and its forms, Curtis felt compelled to reproduce derogatory differentiations that permit the latter to retain its authority.

the expression 'found their way' reveals an anxiety at the moment when modern architecture escaped the exclusive control of western architects ...

Later in the book, in the introduction to Chapter 31, entitled 'Modernity, Tradition and Identity in the Developing World', Curtis says that:

> it was not until the 1940s and 1950s that modern forms had any appreciable impact on the 'less developed' countries, and these forms were usually *lacking in the poetry and depth of meaning* of the masterworks of the modern movement (Curtis 2000: 567; my emphasis).

Here Curtis accuses non-western architects of a lack of sensitivity, their inability to reproduce, or exceed, the poetry and meaning that can be found in certain European and North American buildings – or buildings built by European and North American architects abroad. Note how, for Curtis, architectural 'meaning' is inherent in the buildings themselves: the masterworks of the modern movement are meaningful and poetic because they are modern. As discussed above, superiority is assigned a priori, not justification seems to be necessary. Subsequently, Curtis calls into question the competence of non-western architects who, according to his judgement, are unable to produce work of the same quality as their European counterparts. Admittedly, towards the end of the book Curtis adopts a less unforgiving terminology in order to admit that some of the architectural explorations carried out by architects in the developing world – he refers here specifically to Mexico, Japan, Brazil, Palestine and South Africa – were 'judicious adjustments of generic features of modernism to the climates, cultures, memories and aspirations of their respective societies' (Curtis 2000: 635). Although he tries to reconcile the bi-polar antagonism he had posited earlier in his book, the hierarchical structures are not dismantled. Curtis always refers to non-western architectures in a terminology that stresses either dependency or secondariness (adjustments in this case). Surprisingly, Curtis' derogatory rhetoric has generated little debate in architectural circles; that is, amongst practising architects, architectural academics and students.

This discussion of Curtis's inscription of non-western architectures in the history of the modern movement sheds light on an important absence in architectural debates. Despite an increasing literature influenced by postcolonial theory and, also, in spite of feminist critiques – undoubtedly the two most significant contesting groups – the history of modern architecture, as presented by Curtis, is largely taken for granted and has been seldom challenged. Paraphrasing Bhabha, who asks 'what is at stake in the naming of critical theory as western?' (Bhabha 1994: 31), I ask: what is at stake in naming modern architecture as western? According to Bhabha the answer to the question is rather obvious: 'a designation of institutional power and ideological Eurocentrism' (Bhabha 1994: 31). Indeed, this is what Curtis does when he declares that modern architecture is the 'intellectual property' of the west, he designates an institutional power and reaffirms architecture's ideological Eurocentrism. Bhabha, however, expands further his answer to the above-cited question: 'This is a familiar manoeuvre of theoretical knowledge, where, having opened the chasm of cultural difference, a mediator or metaphor of otherness must be found to contain the effects of difference' (Bhabha 1994: 31).

That is why the writing of architectural history reproduces a relation of domination that has not been challenged critically, thoroughly and efficiently in the field of architecture.

This brief passage from Bhabha's essay 'The Commitment to Theory', which opens the book *The Location of Culture*, expresses with clarity the contradiction found in the writing of architectural history. Here Bhabha maintains that the 'Other (the non-western) is cited, quoted, framed, illuminated, encased in the shot/reverse-shot strategy of serial enlightenment' and, so, the 'Other loses its power to signify, to negate, to initiate its historic desire, to establish its own institutional and oppositional discourse' (Bhabha 1994: 31). Thus, the non-western Other must always be understood in its relation to European norms, or, in the case that concerns us, to European and North American modern architectures. That is why the writing of architectural history reproduces a relation of domination that has not been challenged critically, thoroughly and efficiently in the field of architecture. However, this book shows that the methods of critique advanced by Bhabha, and other postcolonial theorists, do provide the grounds to carry out an interventionist reading of architectural history in order to confront the way in which non-western architectures have been inscribed and represented by western academia.

CHAPTER 4

Hybridity

Perhaps no other term has been more powerful and evocative in postcolonial theory than hybridity. The notion of hybridity has appealed to theorists in virtually all disciplinary areas – including anthropology, cultural studies, geography, literature, sociology and, of course, architecture – as a useful vehicle to study the particularities of sociocultural interaction between different groups in circumstances of colonialism and contemporary globalisation. Rather than simply the straight mixture of two or more elements which form a new one, in postcolonial theory, hybridity has multiple connotations. It refers to the site of cultural productivity that emerges on the margins of culture, between cultures. As such, it is a space where cultural elements are continually rearticulated and reconstituted. Hybridity also expresses the process of rearticulation of culture, hybridisation, a process in which cultural elements change in relation to themselves and to one another; they continue to hybridise. Hence, rather than disappearing in a merger, processes of cultural hybridisation perpetuate difference and, indeed, multiply it. As a result, the concept of cultural hybridity has multiple theoretical effects: it helps to dismantle binary systems of cultural analysis; it unsettles the idea that cultures are, or were, once pure and homogeneous; it disrupts the recognition of authority because it illustrates an endless proliferation of cultural difference; it helps to authorise cultural practices which do not correspond exactly to the parameters of hegemonic systems of cultural classification.

Perhaps no other term has been more powerful and evocative in postcolonial theory than hybridity.

However, hybridity also has adverse implications. Amongst the most notable is that hybridity could be considered as a sign of impurity, the result of a mixture or combination which does not have the same status of the 'original'. In fact, this usage of the term hybridity allows for the confirmation of cultural 'originality' and 'purity', because when something is categorised as hybrid the implication is that it is the result of a combination of elements that are not. Such an understanding of hybridity may arise from the meaning it has in biology, where it refers to the mixture of species, for example a mule, the offspring of a male donkey and a female horse. The mule is a fitting example because it exemplifies how this interpretation of hybridity puts an end to the process of hybridisation (mules are almost always infertile). Moreover, it confirms the idea of inferiority, because mules are not the same as horses which are the 'pure' and 'original', hence, the superior predecessor of the mule. Certainly, this understanding of hybridity had very negative effects in the era of colonialism when 'mixed-race' peoples were heavily discriminated against and, often, considered dangerous precisely because they were neither white nor black.

For Bhabha, hybridisation is the most powerful sign of cultural productivity.

It is these negative interpretations of hybridity that Bhabha sets out to disprove. As we have seen so far in this book, Bhabha is at pains to show that neither languages, nor cultures, nor identities are static or homogeneous. He resorts to theories of translation and to psychoanalysis in order to prove that languages, cultures and identities are fragmented, heterogeneous and ambivalent. For that reason, Bhabha argues, languages and cultures can never fully mix, at least not in the straightforward biological sense. Yet, he asserts that they do interact constantly. In fact, the perpetuation of cultures depends on their interaction with one another. Cultural hybridisation, then, represents the constant, never-ending process of cultural interaction through which cultures continue to exist. For Bhabha, hybridisation is the most powerful sign of cultural productivity.

In order further to explain the significance of the term hybridity in the work of Bhabha, I will proceed to review a theme that has already been introduced: education in Christianity and the European languages. Subsequently, I will analyse Bhabha's own definitions of the term hybridity. Particular attention will be paid to the way in which Bhabha stresses the ability of the concept of hybridity to disturb claims for cultural authority. Considering the theoretical importance of the concept of hybridity, I will take a short theoretical detour to examine some critiques of Bhabha's work made by some of his most incisive commentators. As in the previous chapter, acknowledgement of the critiques of Bhabha is a means to develop an awareness of the theoretical shortcomings that other thinkers have found in his work. This is not in order to take Bhabha to bits, but in order to move the discussion forward; to develop his ideas further and to ease the transition to architecture. In the final part of this chapter I will examine two ways in which the concept of hybridity has been used in architectural studies. These two examples show how Bhabha's ideas could be used to carry out further studies of contemporary as well as past architectures in the peripheries and also in the centres.

Bhabha's hybridity

Bhabha lays out his ideas about hybridity in an article entitled 'Signs Taken for Wonders: Questions of Ambivalence and Authority under a Tree outside Delhi, May 1817', which was first published in *Critical Enquiry* in 1985 and subsequently included as a chapter in the book *The Location of Culture* (1994); the version to which I will refer here. With Bhabha's characteristic wit, the title of the article pairs up two broad, abstract and seemingly opposite concepts, ambivalence and authority, yet he locates the discussion quite precisely under a tree outside Delhi. At the beginning of the article Bhabha retells a story about the discovery of 'the English book' by a group of Indian nomads. The passage, quoted at length, narrates the discussion between an earlier Indian catechist and some people gathered outside Delhi to discuss a translation of the Gospel into Hindoostanee Tongue (the name given by the Europeans to one of the most common languages spoken in India). The people described themselves as poor and lowly but professed great love for 'that book'. The low social rank of the peasants is important in the reading of this passage because the book, the Gospel, teaches about the equality of mankind. As a result, the peasants used the message of the book in order to develop an indifference to the distinctions of the caste system that made them low. The book also served as an instrument to call into question the tyrannical authority of the Brahmins (the priests of Hinduism and, so, a class of scholars and educators). In general, people welcomed the message of the Gospel, which they considered a superior book. At the same time, they refused to believe that the book taught the religion of the European who ate meat.

... colonialism needs to be considered as a complex intersection of multiple subject positions and historical temporalities ...

Bhabha's reading of this passage reveals a moment of instability when the meaning assigned to the 'English book' is rearticulated in the context of cultural and literary translation. This example is also useful to understand the production of meaning in the realm of the untranslatable (see Chapter 2). The meaning and sociocultural significance of the Bible does not reside in the book itself, it has been constructed historically as part of the western Christian tradition which the book represents. Therefore, the literary translation of the book into another language was insufficient to carry forward its religious meaning and social significance; only language was translated. As such, the book becomes an instance of the condition of in-betweeness that I have described before, because in another language the book is no longer the symbol of Christianity and Englishness. In its displaced position (no longer in English, no longer in Europe), it enters a new system of cultural signification and, ultimately, gains new meanings. The English book provides the grounds for the peasants to question their own traditions and the authority that rules them while, simultaneously, rescinding the authority assigned to the English book in the western tradition. By means of this remarkable historical anecdote Bhabha begins to explain how colonialism needs to be considered as a complex intersection of multiple subject positions and historical temporalities, not simply as a straightforward relationship between two assumed homogeneous constructs: colonised and coloniser.

Subsequently, Bhabha proceeds to explain the processes of construction of colonial authority and the conditions necessary to assert that authority. The example of the English book continues to be fitting, because the English book – the Bible as well as literature in general – formed the primary vehicle for the dissemination of European knowledge, the means by which the savage could be enlightened, turned into a copy, or double, of the European. Such was the purpose of the so-called 'civilising mission': to impart European knowledge to the savage in order to improve it, to render it better, to introduce the colonised to the world of reason and progress. Nowhere is this aspiration better expressed than in Thomas Macaulay's infamous Minute of 1835, from which I shall quote a fragment:

> We have to educate a people who cannot at present be educated by means of their mother-tongue. We must teach them some foreign language. The claims of our own language it is hardly necessary to recapitulate. It stands pre-eminent even among the languages of the West. It abounds with works of imagination not inferior to the noblest which Greece has bequeathed to us; with models of every species of eloquence; with historical composition, which, considered merely as narratives, have seldom been surpassed, and which, considered as vehicles of ethical and political instruction, have never been equalled; with just and lively representations of human life and human nature; with the most profound speculations on metaphysics, morals, government, jurisprudence, trade; with full and correct information respecting every experimental science which tends to preserve the health, to increase the comfort, or to expand the intellect of man. Whoever knows that language has ready access to all the vast intellectual wealth which all the wisest nations of the earth have created and hoarded in the course of ninety generations (Macaulay 1835: 349–50).

As is clear from this passage, Macaulay believes that colonised subjects can only achieve progress under the auspices of the English language – the only language that can provide the Indians access to the culture and knowledge of Europe (the west) going back to Greece. Precisely because Macaulay assumes that Indians cannot be educated – in fact, have no 'intellectual wealth' of their own – due to an innate lack in their own language, this passage justifies English colonial rule and education. As explained in the previous two chapters, since colonised subjects, their culture and cultural products, were considered to be inferior (a priori) and backward (they are 'behind' from the point of view of western linear historicity), it is the responsibility of the superior European to bring development to the savage – to civilise them. In other words, Macaulay's Minute gives voice to the European 'burden', that of civilising a people who have no culture; of educating the non-western subject to become European. Based on the ideas of the Enlightenment – progress and rationalism – discrimination is merely a means to bring betterment to the savage. In other words, Enlightenment ideas justify both colonialism and disavowal. At the same time, Macaulay's Minute occludes the damage it causes to the colonised culture. Not only does it conceal the physical damage caused by conquest and colonisation but, more importantly, it obscures the destruction of a culture by means of the imposition of another.

Based on the ideas of the Enlightenment – progress and rationalism – discrimination is merely a means to bring betterment to the savage.

Aware of the magnitude of the task, Macaulay's Minute elsewhere betrays a certain impotence about the possibility that the reproduction of European culture in all colonial subjects and locations might ever be fulfilled. However, for the sake of argument, let us suppose that such a vast project were realised, that colonised peoples were all transformed into Europeans by means of language, religion and education. As explained in previous chapters, if this happened, there would no longer be the duality required for the coloniser to claim cultural authority and to govern the colonised. Consequently, the idea of the 'enlightened' savage becomes a threat to the coloniser: it presents the possibility, even if only distant, that colonised subjects may acquire the same cultural status as the coloniser, or that education in the field of freedom might lead to uprising (as, indeed, happened). In order to counteract the anxiety caused by the potential loss of authority, colonised doubles need to be disavowed for the coloniser to make certain its authority. In this sense, the intent of the civilising mission is also split between its narcissistic desire and the fear of its realisation (see Chapter 3).

In his essay 'Of Mimicry and Man', also included in the book *The Location of Culture*, Bhabha develops an argument about the purpose and procedures of the civilising mission under the concept of mimicry. For Bhabha, 'colonial mimicry is the desire for a reformed, recognisable Other, *as a subject of difference that is almost the same, but not quite*' (Bhabha 1994: 86). In order to explain this evocative statement we can return to another aspect of Macaulay's Minute (quoted in Chapter 2) where he insists that the purpose of the civilising mission was to form a class of interpreters, persons who are Indian in blood and colour but English in taste, morals and intellect. Indeed, a class of people who are 'almost the same but not quite'.

These famous words by Bhabha can also be taken to embody his concept of hybridity in the sense that it represents cultural designations – the English book or a class of people – whose position within the binary structure of colonial representation is imprecise; their difference prevents exact identification or classification within any prior culture (or system of signification). That is why Bhabha further defines mimicry as 'the sign of the inappropriate ... a difference or recalcitrance which coheres the dominant strategic function of colonial power, intensifies surveillance and poses an imminent threat to both "normalised" knowledges and disciplinary powers' (Bhabha 1994: 86). It transpires that mimicry, as a representation of the coloniser's strategy to create a 'double' of itself in the colonised, functions as the very sign by which this 'almost the same' is recognised as 'not quite' and, so, as 'inappropriate'. The key point in Bhabha's argument is that difference or, rather, the means used to differentiate the 'inappropriate', turns against the rational structures that grant authority to the coloniser's 'appropriate' culture, their normalised knowledges and disciplinary powers. Hence, the inappropriate turns into a threat. Such a threat arises from both the impossibility to situate, or to classify, the 'inappropriate' within the rational structures of the Enlightenment and from the difficulty rationally to justify the disavowal of a class of people the coloniser has created as a copy of itself.

> If the concept of mimicry refers to the process of doubling (the purpose of the civilising mission), hybridity represents the cultural products of such an imbalanced and contradictory (ambivalent) process.

The purpose of this abbreviated discussion on Bhabha's concept of mimicry is to contextualise the emergence of hybridity as a result of the mimetic colonial strategy. In fact, close inspection of the two essays ('Of Mimicry and Man' and 'Signs Taken for Wonders') reveals that both concepts are theorised along the same methodological framework, translation and ambivalence, and both have the same purpose: to unsettle the basis upon which claims for colonial authority are made. If the concept of mimicry refers to the process of doubling (the purpose of the civilising mission), hybridity represents the cultural products of such an imbalanced and contradictory (ambivalent) process. In Bhabha's own words:

> Produced through a strategy of disavowal, the reference of discrimination is always to a process of splitting as the condition of subjection: a discrimination between the mother culture and its bastards, the self and its doubles, where the trace of what is disavowed is not repressed but repeated as something different – a mutation, a hybrid (Bhabha 1994: 111).

If colonial mimicry is the desire for a subject who is 'almost the same but not quite', then hybridity is the term Bhabha uses to represent those discriminated identities which signal the ambivalence of the colonial project. An ambivalence that can be seen in the repetition of cultural signs that emerge already different, as 'mutations' rather than the real thing – a difference that is a sign of inferiority. In other words, the practical realisation of the colonial project depends on the production of differentiations, or 'identity effects' as Bhabha calls them (Black, effeminate or underdeveloped), which allow both the narrative construction of cultural superiority and the exercise of power.

After describing the conditions in which colonial hybridity emerges, it seems appropriate to examine Bhabha's own definitions of the term, of which I have selected two; the most extensive explanations that he has produced. The first instance is taken from the essay 'Signs Taken for Wonders'. Here Bhabha defines hybridity thus:

> *Hybridity is* the sign of the productivity of colonial power, its shifting forces and fixities; it is the name for the strategic reversal of the process of domination through disavowal (that is, the production of discriminatory identities that secure the 'pure' and original identity of authority). *Hybridity is* the revaluation of the assumption of colonial identity through the repetition of discriminatory identity effects. It displays the necessary deformation and displacement of all sites of discrimination and domination. It unsettles the narcissistic demands of colonial power but reimplicates its identifications in strategies of subversion that turn the gaze of the discriminated back upon the eye of power (Bhabha 1994: 112).

It is clear that Bhabha assigns an immense critical capacity to the term hybridity. Yet, in order to understand its multiple implications, it is necessary to unpack, as it were, some of the ideas embedded in this quotation. In the first place we note that the primary implication of the term hybridity changes from being derogatory to being subversive. As noted above, the narrative of cultural superiority works both to posit the colonised subject as inferior 'hybrid' and to unify, homogenise, the notion of an English (or European) identity. Macaulay's Minute, for example, assumes that *all* English speakers are well-educated and the possessors of scientific, cultural and mathematical knowledge shared by a community imagined as an 'us'. In Bhabha's view, however, hybridity is no longer the sign of the 'inappropriate' which implies the existence of a prior, original and pure culture, but a sign of cultural productivity which undermines the ideas of both originality and purity. In other words, Bhabha uses hybridity as a theoretical means to overturn the assumption that the coloniser (the 'mother culture') is a homogeneous cultural construct uninhabited by differences. So, Bhabha makes English (European) culture heterogeneous in the same way that he had argued that the colonised is not a homogeneous cultural assemblage. Through heterogeneity Bhabha opens up a field where the authoritarian text, or the authoritative claim, can be misinterpreted, misread, misappropriated by a multiplicity of peoples (as in the example of the Bible discussed above). In other words, by dismantling homogeneity Bhabha stresses the irreducibility of cultural difference. In turn, cultural difference disturbs both the recognition and the exercise of authority, colonial or otherwise. As Bhabha points out, the 'presence of authority is properly established through the non-existence of private judgment and the exclusion of reasons in conflict with the authoritative reason' (Bhabha 1994: 112). So, if the condition of homogeneity necessary for the construction of authority is disproved, its exercise becomes problematic. As a result hybridity gains another dimension under Bhabha's hand: it is not simply a representational term which refers to an inappropriate colonised subject, it is also a seditious concept that signals the 'strategic reversal of the process of domination through disavowal'.

> ... hybridity is no longer the sign of the 'inappropriate' which implies the existence of a prior, original and pure culture, but a sign of cultural productivity which undermines the ideas of both originality and purity.

Another issue that arises from the previous quotation is that hybridity does not refer to a straightforward combination of elements (which is what many architects have taken hybridity to be) but to a process, the productivity of colonial power, and to the conflicts and tensions present in that process. Going back to the passage about the discovery of the English book by a group of peasants outside Delhi, hybridity appears at the moment when the English book, as interpreted by the peasants, ceases to be a symbol of English national authority and becomes a sign of difference; the creative difference of colonised subjects. Hence, hybridity does not lie *only* in the fact that the Bible has been translated into another language, handwritten by various scribes, bound in various forms. It is *also* the fact that the book becomes a sign of cultural productivity and contention at the margins between cultures, or systems of meaning. That is why Bhabha posits the idea that the hybrid object evades epistemological classification. In other words, the hybrid is not a specific form of knowledge nor is it a product, or by-product, that results from the straight fusion of two (or more) elements; or, as Bhabha explains, the hybrid 'is not a third term that resolves the tension between cultures, or the two scenes of the book, in a dialectical play of recognition' (Bhabha 1994: 113–14). Instead, the hybrid is both the cause and the result of the tension between cultures.

> ... the hybrid is not a specific form of knowledge nor is it a product, or by-product, that results from the straight fusion of two (or more) elements ...

In the final sentence of his definition Bhabha indicates that hybridisation causes a reversal that turns the gaze of the discriminated back to the eyes of power. He argues that the ambivalence inherent in the strategy of constructing authority through a repetition (of the coloniser), that is itself a difference, backfires, as it were, causing the emergence of a space of contestation. In his own words, 'the ambivalence at the source of traditional discourses on authority enables a form of subversion, founded on the undecidability that turns the discursive conditions of dominance into the grounds of intervention' (Bhabha 1994: 112). Examples of the seditious effect of hybridity are the re-signification of the English book cited by Bhabha, or the case of the Guajiro Indians in northern South America, discussed in Chapter 2; a case which shows how the Guajiro adopted the language of the Spanish ruler and the system of trade that they imposed, but transformed them in such a way that they could use those traits to their own economic advantage and protection. The Guajiro reversed the conditions of domination to the extent that they were never fully conquered (politically and militarily) nor were they fully converted to Christianity – of course they were unable to return to the conditions of life prior to colonisation, they remain hybrid. However, their creative hybridity is not a sign of inappropriateness but a strategy of resistance, the means through which they retain various aspects of their cultural identity instead of being fully absorbed (disappearing) into the culture of the Spanish.

Let us now draw on another explanation of hybridity as a way to elucidate further the implications of the term. Bhabha advanced these complementary ideas in an article entitled 'Cultures in Between' published in the journal *Artforum* in 1993. The following explanation may help to clarify some of the points raised in the previous definition while introducing other interesting issues:

> **In my work I have developed the concept of hybridity to describe the *construction* of cultural authority within conditions of political antagonism and inequity. Strategies of *hybridisation* reveal an estranging movement in the 'authoritative', even authoritarian inscription of the cultural sign. At the point at which the precept attempts to objectify itself as a generalised knowledge or a normalising, hegemonic practice, the hybrid strategy or discourse opens up a space of *negotiation* where power is *unequal* but its articulation may be *equivocal*. Such negotiation is neither assimilation nor collaboration. It makes possible the emergence of an 'interstitial' agency that refuses the binary representation of social antagonism. Hybrid agencies find their voice in a dialectic that does not seek cultural supremacy or sovereignty. They deploy the partial culture from where they emerge to construct visions of community, and versions of historic memory, that give narrative form to the minority positions they occupy (Bhabha 1993: 167–214; my emphasis).**

Here Bhabha gives us a summary of his understanding of hybridity. The article was published eight years after the essay 'Signs Taken for Wonders' first appeared in 1985. These eight years may have given Bhabha an opportunity to reflect on his initial ideas and to develop a clearer, more eloquent explanation of hybridity. The main issues remain unchanged. In the first part, Bhabha talks about the strategies of constructing authority in conditions of political antagonism. He stresses once more that the plurality and dynamism of transcultural relations prevent classification within binary systems of social antagonism (colonised/coloniser, minority/majority, and so on). Subsequently, he reminds us that the ambivalence inherent in such strategies of authority construction opens up an interstitial space of negotiation where inequality is not eliminated but the articulation of power is equivocal, contradictory. It is worth insisting on this point because the strategic reversal to which Bhabha refers in his writings is often misunderstood as an

The unequal distribution of power persists as an inevitable characteristic of cultural relations (past and present), but the ambivalence inherent in both the discourse and the exercise of power complicates its readability …

erasure of existing structures of power, or power itself. That is not the case. The unequal distribution of power persists as an inevitable characteristic of cultural relations (past and present), but the ambivalence inherent in both the discourse and the exercise of power complicates its readability; ambivalence makes hierarchical structures difficult to perceive and difficult to maintain. Bhabha, then, concludes his definition by suggesting that hybridity, as a cultural condition, allows minority positions to deploy the partial culture from where they emerge and, so, to construct their own visions of community and their own version of historic memory. In other words, Bhabha concludes his definition by assigning agency to minority peoples to contest hegemonic power. Since we have already touched on most of the above-mentioned points, I will concentrate on the last two issues, the partiality of culture and the emergence of minorities, two aspects that have not been addressed yet.

The representation of culture as partial is a key issue in Bhabha's understanding of the term hybridity. To explain the notion of partial culture, or the 'partialising process of hybridity', Bhabha draws on the idea of the English book again. He says that although the book retains its presence, it no longer represents the intended meaning of either the Gospel or English nationalism; therefore its presence in the context of the colony is only partial. 'Deprived of their full presence', Bhabha argues:

> the knowledges of cultural authority may be articulated with forms of 'native' knowledges or faced with those discriminated subjects that they must rule but cannot longer represent. This may lead, as in the case of the natives outside Delhi, to questions of authority that the authorities – the Bible included – cannot answer (Bhabha 1994: 115).

The authorities are unable to answer the questions asked by the natives because they do not correspond with the knowledge (or epistemological system) of the coloniser. The questions relate to the English book but are asked from another position outside the system of the book (European culture). Partiality, then, reinforces Bhabha's idea that cultures are never complete. Cultures are always partial, always in the process of being made and that process takes place mostly on the edges of culture. Consequently, cultures (and the knowledges they contain and represent) cannot be totalised or signified via all-encompassing figures like Englishness or Europeaness. Ultimately, the partialising process of hybridity terrorises the coloniser because it reveals that there are always positions in conflict with authority and, also, that the very enunciation of authoritarian narratives is marked by the trace of difference.

This leads us to the agency of minorities and the partial cultures which they represent. For Bhabha hybridity opens up a site for the emergence of minority positions which have always contributed to the perennial hybridisation of cultures but whose presence in the national context may have been silently repressed, i.e. previously colonised peoples, women, migrants, gays, lesbians and so forth. Certainly the idea that hybridisation opens up a 'space' of cultural negotiation where power continues to be unequal but its articulation is equivocal creates favourable theoretical conditions for minority voices to emerge and to be heard.

Hybridity and hybridisation are important concepts to address questions about minorities because they do not aim to reduce but to maintain difference as an inherent characteristic of all cultures ...

Hybridity and hybridisation are important concepts to address questions about minorities because they do not aim to reduce but to maintain difference as an inherent characteristic of all cultures and, so, it permits the theorisation of minority positions as active participants in the continued production of cultures (see Chapter 6).

At the beginning of this chapter, I commented on the wit of Bhabha's title 'Signs Taken for Wonders: Questions of Ambivalence and Authority under a Tree outside Delhi, May 1817'. The title presents a suggestive mixture of seemingly disparate components. First we have a reference to broad and abstract concepts such as ambivalence and authority. There is also a setting, a suggested geographical location, but it is rather imprecise: a tree outside Delhi. However, the title ends with great chronological precision 'May 1817'. Indeed, the title seems to encapsulate the multidimensionality of the concept the essay deals with: hybridity. Like Bhabha's title, the concept of hybridity appears to have multiple implications and theoretical applications. Some are abstract, intangible, while others are more precise and objective. Hybridity has been defined as a site, an imprecise position on the margins between cultures, a location somewhat similar to that tree outside Delhi in Bhabha's title; a site somewhere not quite in the centre of Delhi but nowhere precisely, yet a place where a lot of things are clearly happening. Hybridity has also been defined as a constant process through which cultures and cultural elements are rearticulated and gain renewed meanings, as the questions to which Bhabha's evocative title refers, questions which reveal the ambivalence of the book as a sign of authority. Hybridity has also been defined as a synthetic product, a result (a book, a building) which gives a sense of finitude and precision, almost like a date: May 1817. However, these different dimensions, or understandings of hybridity, cannot be taken separately because for those finite synthetic produces to continue to be produced there must be a place (a site) where synthesis never occurs because elements remain apart in a state of constant hybridisation. If I may stretch the point, it is the tense coexistence between those elements in the in-between space of hybridisation which guarantees that culturally hybrid products will always continue to be produced. And it is precisely the idea of cultural productivity, the proliferation of difference embodied in the notion of hybridisation, which obstructs the exercise of power. It is also the versatility of the notion of hybridity that has appealed to architects, architectural theorists and historians. In its most basic form, it allows for the theorisation of buildings and cities which combine different forms, materials and ornaments. However, Bhabha's concept of hybridity, as has been explained so far, provides myriad opportunities to engage with issues beyond form and image and, so, to exceed the limitations of existing methods of architectural critique connecting architecture with a broader set of socio-political and cultural issues which are often ignored.

Critiques of hybridity

Although hybridity is one of Bhabha's most powerful and evocative terms, it has also generated a great deal of controversy. Of all the concepts that he uses in his critiques of colonial discourse (mimicry, performative and, even, his use of psychoanalytic ambivalence), hybridity is the most heavily criticised. For that reason, before exploring the potential that this concept offers for the development of architectural debates, it is useful to address some of the main critiques that have been made of Bhabha's usage of hybridity. Not only will a brief analysis of such critiques help to situate the work of Bhabha in a wider context, it may also help to understand his ideas better in the light of his critics' interpretations.

Although hybridity is one of Bhabha's most powerful and evocative terms, it has also generated a great deal of controversy.

Let us start with Jane M. Jacobs, author of *Edge of Empire: Postcolonialism and the City*, whose work is particularly relevant in the field of architecture. Jacobs criticises Bhabha because his questioning of colonial authority seems to focus exclusively on the internal ambivalence of colonialism – its discourse and discriminatory strategies – rather than on the colonised people as the agents of subversion. In other words, it is the unveiling of colonialism's own internal flaws which undermines the coloniser's claim for authority while colonised people appear to be devoid of subversive agency. In Jacobs' own words, 'this is because

Bhabha's main concern is with the field of colonial discourses rather than anti-colonial discourses and formations' (Jacobs 1996: 28). Considering that Bhabha himself maintains that 'resistance is not necessarily an oppositional act of political intention', but 'the effect of an ambivalence produced within the rules of recognition of dominating discourses as they articulate the signs of cultural difference and reimplicate them within the differential relations of colonial power' (Bhabha 1994: 110–11), Jacobs' criticism appears to be somewhat valid. In other words, Jacobs accuses Bhabha of failing to do justice to the politics of contestation manifested in counter-colonial movements and other forms of collective insurgency.

Likewise, Robert Young, one of Bhabha's most incisive critics, has questioned acutely the way in which Bhabha uses the very terms hybridity and hybridisation without carefully considering their theoretical tradition or the different cultural situations these terms are used to describe. Young criticises explicitly the fact that Bhabha uses a variety of terms including hybridity and mimicry, amongst others, in order to describe the ambivalent conditions of colonial discourse. Yet, he makes 'no reference to the historical provenance of the theoretical material from which such concepts are drawn, or to the theoretical narrative of Bhabha's own work, or to that of the cultures to which they are addressed' (Young 1994: 186). Although very compact, Young's critique deals with three important aspects. First is Bhabha's failure to refer to the historical provenance of the terminology he employs. In Young's opinion, this is problematic because such terms are loaded with socio-political significance in relation to the contexts – theoretical and geographical – in which they have been previously used. For example, hybridity has been used in linguistics to describe words that combine a prefix or suffix from one language with a stem from another. It has also been used in biology to describe the process of mixing different species together usually in order to increase productivity or to make a certain species resistant to adverse natural conditions. However, it is the way in which the concept of hybridity was used in the nineteenth century as the very means of colonial discrimination on grounds of ethnicity and race which Young finds most troubling. In this context, Young maintains, the notion of hybridity stands as the very opposite of Bhabha's intended meaning. Young does not dismiss Bhabha's argument entirely, he actually identifies with Bhabha's proposition that hybridisation has an unsettling effect on colonial discourse, but highlights the need for the historical contextualisation of the chosen terminology. Otherwise, in Young's view, the theoretical ability that Bhabha assigns to those terms would be reduced due to a lack of semantic (and historical) precision.

The second point that arises from Young's critique refers to the wide choice of terms used by Bhabha in order to reiterate his critique of colonial discourse. Young suggests that Bhabha shifts constantly from one term to another and, yet, it appears that each term accomplishes the same theoretical objective: to question the authority of colonial discourse. As Young puts it:

> On each occasion Bhabha seems to imply through this timeless characterisation that the concept in question constitutes the condition of colonial discourse itself and would hold good for all historical periods and contexts – so it comes as something of a surprise when it is subsequently replaced by the next one, as, for example, when psychoanalysis suddenly disappears in favour of Bakhtinian hybridisation, only itself to disappear entirely in the next article [where] psychoanalysis returns, but this time as paranoia. It is as if theoretical elaboration itself becomes a kind of narrative of the colonial condition. Inevitably, of course, different conceptualisations produce different emphases – but the absence of any articulation of the relation between them remains troubling (Young 1994: 186–7).

In spite of his criticism, Young admits that 'it is possible to detect a certain schema being worked through, although it is almost unrecognisable in its dissimilitude' (Young 1994: 187). He then argues, in favour of Bhabha, that the complexity of colonialism is such that no single concept would be sufficient fully to describe it or to advance a critique of it. Hence, it is necessary to resort to a variety of concepts, each of which relates to specific issues and moments in history. Clearly, Young's demand is for Bhabha to elaborate further on the emphases produced by each one of the concepts that he employs and to articulate their usage in order to create a more coherent and effective critical narrative.

The third issue raised by Young refers to Bhabha's lack of reference to the cultures that he addresses in his text. This is, arguably, one of the most problematic aspects of Bhabha's work, one that has recurred in the work of some of his most acute critics, for example Bill Ashcroft, Jane Jacobs, Ania Loomba and Benita Parry. They suggest that Bhabha applies his theories to a variety of contexts indistinctively of their specific circumstances: historical, social, political and economic. Certainly Bhabha moves rather swiftly from colonial India to present-day New York via numerous mentions of countries such as Algeria and Sri Lanka in different historical periods. As a result, Bhabha's notion of hybridity becomes problematic because it generalises on the characteristics of the colonial situation and the colonial hybrid. In other words, what is presented as a strategy of differentiation ironically acquires universal and homogeneous characteristics common to the colonial relation per se.

This is a delicate issue because the purpose of the notion of hybridity in postcolonial discourse is not to affirm that all cultures are hybrid but, on the contrary, that all hybrids are different.

This is a delicate issue because the purpose of the notion of hybridity in postcolonial discourse is not to affirm that all cultures are hybrid but, on the contrary, that all hybrids are different. The concept of postcolonial hybridity refers to the particular characteristics of the relation between the colonised and the coloniser – and any other participating sociocultural groups involved in the colonial relation, i.e. slaves, merchants, etc. – in specific geographical locations at precise moments in history. Clearly, relations between colonised and colonisers vary greatly from seventeenth-century Bolivia to South East Asia in the nineteenth century, or from India in the eighteenth century to Algeria in the early twentieth century. While it is possible to assert that cultural hybridisation occurred in all these places due to colonialism, it is disingenuous to suggest that the conditions in which it happened were alike. The critical efficacy of the concept of postcolonial hybridity lies in the fact that it unveils the very struggle between different groups to retain their identities (though transformed) in conditions of inequality and inequity – which is, indeed, what Bhabha says, although in a more extended manner.

Criticism, thus, has not led to the demise of hybridity debates but to its theoretical optimisation.

By looking at some of the most penetrating critiques of Bhabha's work, I have intended to shed further light on the implications and the potential of hybridity in postcolonial theory. In fact, the work of critics such as Ania Loomba, Benita Parry, Jan Nederveen Pieterse, Ella Shohat and Robert Young, amongst many others, has contributed to the dissemination of Bhabha's ideas as well as to the development of his critique. Criticism, thus, has not led to the demise of hybridity debates but to its theoretical optimisation. That is why Bhabha's notion of hybridity remains a key concept in postcolonial studies, and, as such, a term which has had a great deal of impact in architectural debates during the past 30 years.

Hybridity as form in architecture

As mentioned above, the notions of hybridity and hybridisation have influenced greatly the development of contemporary architectural debates. These concepts have been particularly useful in discussions about the architectures of previously colonised peoples as well as buildings produced in the so-called developing world. In early discussions, however, architectural theorists and historians approached hybridisation mostly because of its semantic implications: a term that represented a combination of materials, forms, construction techniques and ornamentation. In this sense, hybrid architectures were implicitly set against a background of 'non-hybrid' architectures or against the assumed purity of European architecture whose origins can be traced back to antiquity. Therefore, in this context, hybridity established a hierarchical system which gave authority to European architectures; an authority based upon its anteriority and its alleged homogeneity.

... in this context hybridity established a hierarchical system which gave authority to European architectures ...

An example of this usage of the concept of hybridity is found in the work of Chris Abel who, in his book *Architecture and Identity: Responses to Cultural and Technological Change*, dedicates an entire chapter to the study of architectural hybridity. The chapter, entitled 'Living in a Hybrid World', discusses the fusion of different architectural elements in the context of colonial Malaysia and begins with a brief analysis of the indigenous Malay house. Abel describes carefully the formal characteristics of the Malay house which, in his view, responds not only to local environmental conditions but also to cultural traditions and the social organisation of its inhabitants. After his description of the Malay house, Abel examines, via various case studies, the process of formation of what he calls a 'new' Malaysian architecture.

Abel's first case study is a typical British colonial house located in Georgetown, on the island of Penang. Abel chose this case because it is similar to earlier houses built by British colonisers in Malaysia and, also, because it is a European building type: a villa. Abel proceeds to trace the formal genealogy of the house back to the Italian Renaissance and the villas of Andrea Palladio. As Abel puts it:

> The basic form, architects will easily recognise, is that of Palladio's *villa suburbana*. It is, at one level, a most successful illustration of the manner in which classical architecture travels over time and space. In this case, from Venitia in northern Italy, where Palladio built almost all his villas during the middle sixteenth century, to England by courtesy of Inigo Jones and the English Palladians of the eighteenth century, to this very different locale but still very English suburb on Penang Island in South East Asia (Abel 1997: 153).

After establishing the European lineage of the building, Abel explains how British architects in Malaysia progressively incorporated formal elements taken from the local Malay house into the Italian villa type as a way to adapt the latter to the climatic conditions of the tropics. The basic form remained unchanged – plans, proportions and compositional elements – but some features were altered in order to allow for cross-ventilation and to create protection against both torrential rain and excessive exposure to sunshine. In this way, Abel maintains, the suburban villa built by British architects in Malaysia was 'clearly no longer *of* Italy and England alone or even together, but belongs *to* tropical Malaysia' (Abel 1997: 154; emphasis original).

In his second case study, Abel examines the Chinese-built 'shophouse' as an example of 'the classical language of architecture being used for a non-Western building type and social form' (Abel 1997: 155). In this case, the classical orders and other motifs are appropriated from European architecture but used in different ways; different from the European *original*. Once again, Abel explains how European ornamentation and classical orders were transformed in order to respond both to tropical weather conditions and to the particularities of urban trade in colonial Malaysian towns.

The third case is a government building designed and built by the colonial administration to house its offices. This time, Abel explains, a building originally designed in the classical style by an English architect is subsequently redesigned by his superior in the Public Works Department. The latter had worked in Ceylon and admired the Saracenic architecture of India. Hence he proceeded to combine classical and Islamic forms; a process which, according to Abel, 'produce[d] what is instantly recognised as the characteristic features of the historic government buildings of Kuala Lumpur' (Abel 1997: 158). It thus transpires that, for Abel, the architecture of Malaysia is characterised by the minimal adaptation of European building types and by a combination of classical European and Islamic architectures. However, Abel makes no effort to discuss the role of local pre-colonial architectural traditions in the formation of contemporary Malaysian architecture.

Let us pause momentarily to reflect on Abel's account about the formation of what he calls a *new* Malaysian architecture, a hybrid architecture. First, it is important to note that Abel does not refer to the way in which the Malay house was altered due to the influence of colonialism. He focuses only on the way in which the European villa type was transformed by 'British architects' who introduced into it elements taken from the Malay house. Whatever happened to the indigenous Malay house after the arrival of the European is of no concern to Abel, who seems to be preoccupied exclusively with the effects of cultural interaction on European architectures. Abel fails to explain whether the Malay house was also transformed as a result of colonialism, or whether it subsequently fell into disuse. It follows that what Abel calls the *new* identity of Malaysian architecture is a transformed version of European architectural styles produced by British architects in the colony – colonial subjects and their architectures are completely dismissed.

A similar argument can be developed about the second and third case studies. It is clear that European building types and styles were altered, and combined with others, in order to respond to the specific conditions of Malaysia – not only its climate but also the particularities of trade and the economy brought about by British colonisation. However, there is no reference to the native architectures that these 'hybrid' buildings replaced. Abel is unable to develop an argument in this regard because, in both cases, the new types were the result of impositions brought by colonisation: capitalism (in the case of the shop) and a central

> ... this form of hybridisation recomposes the idea of an original and homogeneous European architectural historicity that unites the English with the Italian Palladians via Inigo Jones.

governing system (in the case of the administrative building). In this case hybridisation is something that happens to European building types and styles. It either does not happen to non-western architectures or their historical hybridisation is unimportant to the historian. Moreover, this form of hybridisation recomposes the idea of an original and homogeneous European architectural historicity that unites the English with the Italian Palladians via Inigo Jones. Simultaneously, it implies that the Malaysian hybrids, although produced by enlightened British architects and engineers, are of a lesser standard than the originals (which were pure). It is possible to identify an instance of colonial mimicry, the desire for a *reformed* native whose architecture is intrinsically inferior and, consequently, needs to be improved through the architectural methods of the coloniser – for, as pointed out above, it is assumed that colonised subjects can only achieve progress under the auspices of European architectures. Abel's representation of Malaysian architectural identity as a transformed version of European building types occludes any prior sign of history. This omission suggests a process of historical erasure: the removal of non-European peoples from their own pre-colonial past during their inscription into the *universal* linear history of the European. It is the construction of a history of subjects who did not have a history – at least, not as part of the linear western historicity – and who could only attain historical subjecthood through the scholarly historicising methods of the coloniser. The ambivalence of such a procedure, we concluded in the last chapter, calls into question the validity of architectural history and the methods used to represent previously colonised people historically. Abel's way of 'constructing' an identity of Malaysian architecture – as if they could not construct their own – perpetuates the hierarchical system that grants authority to European architectural discourses and practices.

In postcolonial discourse, even the simple operation of mixing styles, materials and techniques triggers questions about who does the mixing, in what circumstances does the mixing take place or who documents that mixing historically.

This brief reading of Abel's chapter on architectural hybridity reveals how Bhabha's method of critique opens up doors to challenge the validity and univocality of architectural history. It also reveals the theoretical difficulty of reducing hybridity merely to its ability to describe the mixture of styles, materials and constructions techniques, or to represent architectural identities the same way. The political content inherent in the term hybridity implies that such mixtures are never innocuous acts. In postcolonial discourse, even the simple operation of mixing styles, materials and techniques triggers questions about who does the mixing, in what circumstances does the mixing take place or who documents that mixing historically. Thus, the notion of hybridity emerges as a sophisticated mechanism for the continued study and understanding of architectures beyond mere form.

Representing non-western architectures

More recently, the adoption of postcolonial methods of critique into architecture has broadened the scope of discussions about architectural hybridity. Rather than focusing only on form, the notion of hybridity serves as a vehicle to link architecture with debates in other disciplinary areas and, ultimately, to politicise architectural discourse. In her book *Hybrid Modernities*, for example, Patricia Morton refers to Bhabha's theory as a way to examine how the French represented the people from their colonies at the World Exposition in 1931. Morton starts the book by asking five questions which delineate her entire argument:

> What happened when the colonies were brought to Paris in 1931?
> Were unexpected meanings produced out of the juxtaposition of the colonies to Paris?

> How were the colonies architecturally represented?
> Was the exposition a convincing colonial environment?
> How was meaning generated by an architecture that served the dual function of representing French colonial power while it represented the colonised societies of the empire? (Morton 2000: 9)

Although Morton does not elaborate extensively on Edward Said's notion of Orientalism, these questions certainly set the discussion in that context: the cultural representation of the colonised as the dark, exotic, barbaric and sexualised Other against whom the coloniser constructs itself as superior. This ambivalent relation, Said stresses, is caused by the fact that the exoticised Other exerts an inexplicable attraction to the coloniser, an attraction which generates great fear of losing their authority over it. However, rather than focusing on Said, Morton prefers to adopt Bhabha's usage of hybridity. She sees a theoretical ability in Bhabha's hybridity which facilitates the representation of indeterminacy, lack of resolution and the need to situate any analysis of colonial relations in the context of a permanent – and agonistic – negotiation:

> **The hybrid pavilions embodied the intersection of the colonised's and the coloniser's experience, the 'in-between' that Bhabha identified as postcolonial space. The exposition occupied the middle region of experience where the norms, rules and systems of French colonialism both emerged and broke down, unsustainable because of their internal contradictions (Morton 2000: 14).**

For Morton hybridity is not fixed in the architecture or the image of the pavilions. Hybridity emerges in relation to the larger context of the Exposition as well as in relation to Paris and France as a whole. For Morton hybridity is a site, yet not a physical site, a territory, but an imprecise location where the 'norms, rules and systems of French colonialism' are reinforced and dismantled concurrently. That is why Morton refers to various forms of hybridity which operate simultaneously in the production of colonial authoritarian discourse as well as in the dismantling of the principles of such discourse.

For Morton hybridity is a site, yet not a physical site, a territory, but an imprecise location where the 'norms, rules and systems of French colonialism' are reinforced and dismantled concurrently.

A first form of hybridity is seen in the native pavilions. These were designed by French architects, not by the natives themselves. The exterior image of the native pavilions was designed so that they reproduced precisely the architecture of native peoples in their state prior to colonisation, as if colonisation had never happened. The interior of the pavilions, on the other hand, displayed the didactic exhibits of progress and French civilisation. In other words, on the exterior, native pavilions represented the savage while their interior symbolised the advancement brought by the French. Although it focuses on the material coexistence of culturally specific things – objects, forms, materials, etc. – Morton addresses the implications of such conflictive coexistence beyond the limits of physicality (rather than a merger). The split between exterior and interior represents a form of hybridisation at the scale of the building.

Another instance of hybridity is found in the different kinds of pavilions: those which represent the architecture of the natives and the metropolitan pavilions. The former were built in the native styles of the colonies and the latter were designed following the aesthetic principles of Art Deco. Such a deliberate differentiation between colonised and coloniser, Morton argues, served the purpose of highlighting the backwardness of colonised peoples and the development of France. The tense and politicised coexistence of native and metropolitan pavilions represents another form of hybridity at the scale of the Exposition.

A more dramatic form of hybridity emerged as a result of the need to meet the requirements of building in such an important European capital as Paris. The scale of Parisian buildings required architects to used *beaux-arts* design techniques in order to magnify, or re-scale, native pavilions – to make them

monumental – yet maintaining the original forms, images and decorative elements of each signifying culture. 'This mingling of representational vocabularies', Morton argues, 'brought architecture at the exposition into the dangerous territory of cross-breeding, so terrifying to the colonials' (Morton 2000: 197). Morton elaborates further on the terror caused by cross-breeding, especially when it happened spontaneously rather than under the control of the authorities; architects in this case. Morton explains that there was an increasing fascination with 'exotic' architectures in France at the time of the Exposition. This fascination had materialised in numerous eclectic constructions of *Oriental fashion* throughout Paris, and France in general. Such a fascination with primitive cultures had already had an enormous effect on other aspects of French culture apart from architecture. It generated *bals nègres*, jazz, *negrophilia* and primitivism, all of which were considered to be part of a decadent cosmopolitan culture. The eclectic spontaneous constructions that proliferated in Paris at the time as a result of the fascination for Oriental motifs were considered pastiche and, consequently, were unacknowledged by the organisers of the Exposition. For them, eclectic spontaneous architectures embodied the horror of cultural mixing – whereas the native pavilions designed by French architects following the principles of the *beaux-arts* did not pose a threat to their authority. That is why Morton asserts that the paradox inherent in French colonialism 'is that it produces hybrids even while repudiating them' (Morton 2000: 200–1). This last form of hybridity exceeded the limits of the Exposition, and even of Paris, because it refers to France and Europe in general.

That form of hybridisation unsettles the binary structures of colonial representation because it generates a multiplication of difference rather than the containment of it …

Taken together, the three forms of hybridity that Morton identifies in the Paris Exposition of 1931 cause 'the erasure and blurring of boundaries between races and the dissolution of the codes of difference established by colonialism' (Morton 2000: 200). For Morton, the most significant signs of hybridisation are not found within the confines of the Exposition – in the juxtaposition of scales and architectural styles, not even in the contradiction between the interior and exterior of buildings – but on the appropriation of native motifs by the common French person. That form of hybridisation unsettles the binary structures of colonial representation because it generates a multiplication of difference rather than the containment of it; it becomes the *horror* of colonialism. It is when hybridisation exceeds the control of the authorities (architects) that it becomes a menace. Or, in Bhabha's words,

> the paranoid threat from the hybrid is finally uncontainable because it breaks down the symmetry and duality of self/other, inside/outside. In the productivity of power, the boundaries of authority – its reality effects – are always besieged by 'the other scene' of fixations and phantoms (Bhabha 1994: 116).

In spite of her thorough and provocative depiction of architectural hybridity in the context of the 1931 Paris Exposition, the question arises as to what role did colonised peoples, the natives, play in the production of material, cultural and political hybridity? While Morton is at pains to underline how architectural (and other forms of) hybridity obfuscate the rules of recognition of colonial authority, to use Bhabha's own words, it is also apparent that the various forms of hybridity she refers to are produced by the European: French architects (as well as musicians, painters and, even, the French in general). The natives are excluded from the production of hybridity. Consequently, it could be argued that Morton suffers from the same inconsistency that various critics have found in the work of Bhabha: that the concept of hybridity is useful theoretically to unveil the ambivalence inherent in colonialist discourse but, at the same time, denies any subversive agency to colonised peoples. In other words, Morton renders the natives passive participants who are unable to offer resistance.

Architectural hybridity does not lead *only* to the formation of 'new' synthetic architectures. Instead, hybrid architectures are a testimony of the deep and complicated procedures through which they emerge (social, political, historical, economic).

Regardless of such an imprtant theoretical flaw, Morton uses Bhabha's notion of hybridity in order to discuss issues regarding cultural representation in the context of colonialism as well as its intrinsic and perennial contest for authority. Morton's study of hybridity brings to the fore the unequal distribution of power between colonised and coloniser and, to a lesser extent, the tortuous historical experiences of colonised peoples. In her work, hybridity does not refer only to the mixture of forms, materials or decorative elements but to the socio-political effects of such a mixture. Architectural hybridity does not lead *only* to the formation of 'new' synthetic architectures. Instead, hybrid architectures are a testimony of the deep and complicated procedures through which they emerge (social, political, historical, economic). This approach differs from Abel's monolithic and over-simplified version of architectural hybridity. While Abel's intention appears to reinforce European culture by uprooting colonised cultures from their own pre-colonial past, Morton offers an interpretation of architectural hybridity which attempts to acknowledge their agonistic relation.

CHAPTER 5

The Third Space

The 'Third Space' is another concept in contemporary cultural theory instantly associated with the work of Homi K. Bhabha. Although Bhabha himself does not elaborate extensively on this concept, he presents it as the 'precondition for the articulation of cultural meaning' (Bhabha 1994: 38), thereby situating the Third Space at the centre of his discussions on cultural difference and cultural productivity.

<u>… it could also be argued that the Third Space is an attempt to 'spatialise' the liminal position it represents; in other words, to give a certain tangibility to the in-between space where hybridisation occurs, and from where hybrid designations emerge.</u>

By means of the Third Space, Bhabha joins a group of thinkers who feel uncomfortable with the reductionism of dialectical systems that create a 'politics of polarity', that is, oppositions such as self/other, centre/periphery, colonised/ coloniser, and the like. As discussed in the previous chapter, his idea of hybridity is already an attempt to locate culture, the productivity of culture, in a situation of liminality (in a threshold, or passage, between two positions or more). In his own words, the hybrid is 'neither one nor the other', instead, it is always in-between, where it continually transforms itself according to the dynamics of cultural interaction. In fact, in an interview published in 1990 under the title 'The Third Space', Bhabha equates the Third Space to the concept of hybridity. He says that 'hybridity is the Third Space which enables other positions to emerge' (Bhabha 1990b: 211). This assertion has caused some confusion as to whether the two concepts stand for the same thing. However, it could also be argued that the Third Space is an attempt to 'spatialise' the liminal position it represents; in other words, it gives a certain tangibility to the in-between space where hybridisation occurs, and from where hybrid designations emerge.

Theorising the Third Space

To explain his idea of the Third Space Bhabha resorts to a semiotic account of the disjunction between the subject of a proposition and the subject of enunciation. This may seem complicated but it is, in fact, rather simple. The disjunction between the two subjects has been described by Lacan as the dual position of the 'I' in the process of communication. For Lacan, the pronominal 'I' (the speaker) is split because it occupies a double position in the act of enunciation, it is both the signifier and the grammatical subject of the statement. The two positions, however, do not necessarily correspond. To explain the disparity between these two positions Lacan asks: 'is the place that I occupy as subject of the signifier concentric or eccentric in relation to the place I occupy as subject of the signified?' (Lacan 2006: 430). In answering this question, it is interesting to note, first, that Lacan refers to 'I' as a place one occupies at the moment of speaking. Yet, it is also a place one immediately loses when the other person (the interlocutor) responds to us. So, in a dialogue the position of the 'I' (the speaker) is not fixed, it keeps shifting between the two participants; a situation that becomes exacerbated in a conversation between more than two people. Second, the answer to Lacan's question about the concentricity or eccentricity of the signifying 'I' (me, the one who speaks) in relation the signified (me, the person I refer to) is: both. As Lacan himself puts it,

> the point is not to know whether I speak of myself in a way that conforms to what I am, but rather to know whether, when I speak of myself, I am the same as the self of whom I speak (Lacan 2006: 430).

According to Lacan, then, the speaking subject splits at the very moment of enunciation (when one speaks). A division occurs between the 'I', the speaker who says the sentence, and 'I' as the figure (or subject) to which the speaker refers in the sentence. While the speaking 'I' is unmistakable, the second 'I' is not, because the pronoun is insufficient fully to account for what I am, and represent, and stand for. This may sound difficult but it is not. Bear with me. Let us use a classic example here in order to make this seemingly confusing argument clearer. In the sentence 'I love you', there is the 'I' who says the sentence and so declares his/her love to someone else; that 'I' is unmistakable because it is the person speaking. On the hand, there is the 'I' who loves, certainly a much more complex entity than the word 'I' can account for, because the 'I' who loves includes the entire set of feelings, fears, desires, traumas, and so on, which makes the self – and which exceeds the representational capacity of the pronoun. As a result of this split, the singular implication of the pronoun 'I' is turned into a plural. Consequently, the shifting positions of the 'I' in a conversation and the splitting that occurs at the moment of enunciation complicate the process of communication: it is no longer the simple and straightforward exchange between the 'I' and the 'you' but a process of negotiation, contestation and rearticulation where meaning is both produced and lost. The loss or production of meaning occurs in the acts of speaking and interpreting the spoken statement. That is why Bhabha argues that:

> **The production of meaning requires that these two places [the I and the you] be mobilised in the passage through a Third Space, which represents both the general conditions of language and the specific implication of the utterance in a performative and institutional strategy of which it cannot 'in itself' be conscious. What this unconscious relation introduces is an ambivalence in the act of interpretation (Bhabha 1994: 36).**

It is important to note that Bhabha does not intend to advance a definition of the Third Space through this semiotic analogy or through psychoanalysis. He uses this example in order to 'dramatise' the inherent linguistic difference that informs all cultural performances. Like Lacan, Bhabha spatialises the 'I' and the 'you', which are considered to be places. Then, Bhabha proceeds to talk about the Third Space as a passage. As such, the passage is a double figure which stands for language as a system (the general conditions of language) and as speech (performance). Indeed, the figure could also be understood as the very passage from language to speech (the unrestricted performance of language by people). All the above-mentioned characteristics of language and communication make content (the meaning of the message) ambivalent before it is received by the interlocutor, or interlocutors. That is why Bhabha goes on to assert that:

> The intervention of the Third Space of enunciation, which makes the structure of meaning and reference an ambivalent process, destroys this mirror of representation in which cultural knowledge is customarily revealed as an integrated, open, expanding code. Such an intervention quite properly challenges our sense of historical identity of culture as a homogenising, unifying force, authenticated by the originary Past, kept alive in the national tradition of People. In other words, the disruptive temporality of enunciation displaces the narrative of the western nation which Benedict Anderson so perceptively describes as being written in homogenous serial time (Bhabha 1994: 37).

Let us return to an example that we have already examined in order to expand on this idea: Macaulay's Minute. In the previous chapter we saw that Macaulay assumes that *all* English speakers are well-educated, the possessors of scientific, cultural and mathematical knowledge and, so, he represents the English, or Englishness, as an integrated, homogeneous cultural construct validated by its classical provenance. However, his Minute occludes the fact that not all English people are educated and possess the wealth of knowledge that Macaulay alleges links the English with the Ancient Greeks. That is why Bhabha asserts that:

> it is only when we understand that all cultural statements are constructed in this contradictory and ambivalent [third] space of enunciation, that we begin to understand why hierarchical claims to the inherent originality or 'purity' of cultures are untenable, even before we resort to empirical historical instances that demonstrate their hybridity (Bhabha 1994: 37).

The idea of the Third Space emphasises the rhetoric of ambivalence which Bhabha uses throughout his entire body of work in order to undermine not simply the authority of colonial discourse but, also, the means through which it is exercised. By locating ambivalence at the very moment of enunciation, even before the statement is received and subject to interpretation, it is always-already split; its authority is already questionable.

Spatialising the Third Space

It is clear that Bhabha is attracted to the spatial connotations inherent in the 'Third Space' as a critical term. Paradoxically, many of Bhabha's commentators indicate that the Third Space is not actually a space that can be entered or left; at least not in the physical way architects understand it. Bhabha himself asserts that the Third Space is unrepresentable. However, in the foreword to the book *Communicating in the Third Space*, one of the most recent publications on the topic, Bhabha illustrates his idea of the Third Space through a simile:

> Across the threshold of terror and genocide that joins the twentieth century to the twenty-first, Truth Commissions provide a dialogical Third Space committed to democratic processes of political transition and ethical transformation in societies made wretched by violence and retribution. The traditional grass mat, the *gacaca*, that has provided a name and place for local practices of post-genocide adjudication in Rwanda – the *Gacaca* Courts – also qualifies as a Third Space. The *gacaca* is not simply a space of confession, not is it principally a space of confrontation and guilt. It is a place and a time that exists in-between the violent and the violated, the accused and the accuser, allegation and admission (Bhabha 2009: x).

It is a place and a time that exists in-between the violent and the violated, the accused and the accuser, allegation and admission.

Quite clearly, Bhabha gives physicality to the notion of the Third Space by comparing it with a courtroom, although it is a particular kind of courtroom with geographical, historical and political specificity. However, he seems reluctant to limit the theoretical possibilities of the Third Space to the confines of a physical enclosure. For that reason, after this simile, he develops an alternative – or complementary – argument through a metaphor. He invites us to consider an incident in Joseph Conrad's *Heart of Darkness*, a book regularly mentioned in the field of postcolonial studies. The incident involves Marlow, the protagonist of the novel, who, aware of the asymmetries of warfare and hesitant about considering the natives as his enemies, approaches a moribund man whose face Marlow suddenly finds near his hand. Next Marlow sees that the black man had a bit of white worsted tied about his neck, 'why? Where did he get it? Was it a badge – an ornament – a charm – a propitiatory act?' In citing this incident and Marlow's questions, here, too, the Third Space is given physicality. However, this time, the Third Space is not considered to be an actual space, but an object, a piece of thread tied around the neck of a dying native:

> Somewhere between the two, Marlow enters a Third Space. He is now engaged in a translational temporality in which the 'sign' of the white worsted from beyond the seas, is an object of intention that has lost its mode of intention in the colonial space and vice versa. The familiar origin of the worsted as a commodity of colonial trade passes through an estranging realm of untranslatability in the heart of darkness, and emerges ready to be raised anew and at other points in time (Bhabha 2009: xii).

The thread opens up an area of irresolution between Marlow and the native, a Third Space where the brief relation between the two characters is expanded and made complex, rather than resolved. The simple dichotomy between native and metropolitan is suddenly insufficient to explicate how the native obtained the worsted or why he was wearing it. An interesting aspect of Bhabha's discussion is the reference to a worsted, which is the name for a yarn (and the fabric made from this yarn) traditionally made in the village of Worstead, in Norfolk, England. Worstead became a prosperous town in the twelfth century when immigrant weavers from Flanders were encouraged to settle in England

and established their spinning and weaving industry there. Thus, a remarkable symbol of Englishness is produced by Flemish migrants, a fact which accentuates Bhabha's general notion of the Third Space as a passage, a space of negotiation, contestation and rearticulation, where a yarn produced by Flemish migrants in Norfolk becomes a symbol of Englishness which gains new meanings when tied around the neck a moribund man from beyond the seas.

The question that arises here is whether the native also entered the Third Space, or whether it was only Marlow, the dominant, for whom the thread became an open question. Bhabha speaks of the native as 'an agent caught in the living flux of language and action' but it is unclear whether that agency is assigned by the narrator (Marlow) or by Bhabha. What is clear is that, through the *gacaca* simile and the worsted metaphor, Bhabha reaffirms his interpretation of the Third Space as a liminal site between contending and contradictory positions. Not a space of resolution, but one of continuous negotiation. 'The Third Space', Bhabha asserts, 'is a challenge to the limits of the self in the act of reaching out to what is liminal in the historic experience, and in the cultural representation of other peoples, times, languages, texts' (Bhabha 2009: xiii).

Bhabha reaffirms his interpretation of the Third Space as a liminal site between contending and contradictory positions. Not a space of resolution, but one of continuous negotiation.

It is also interesting that in his later writings Bhabha moves away from the highly theoretical, somewhat abstract definitions of the Third Space given in his early articles. As is clear in the last two quotations, the discussion gravitates around specific events, both historical and literary, which reveal a markedly human dimension. It could be argued that Bhabha has developed his conceptualisation of the Third Space retrospectively, or that he now can post-rationalise some of his original ideas in the light of other theories of 'thirdness'.

Third Space and architecture

In fact, Third Space scholars like Edward Soja argue that the term was inspired by Henri Lefebvre in his book *The Production of Space*. As a geographer, Soja is interested in exceeding the limitations found in the dualism between first and second space. First space is generally understood as all forms of direct spatial experience, that is, those which can be measured and represented cartographically. Second space, on the other hand, 'refers to the spatial representations, cognitive processes as well as modes of construction, which give rise to the birth of geographical imaginations' (Soja 2009: 51). Geographical thinking, Soja argues, was defined by this limiting dualism until the 1990s, when Lefebvre's work was first translated into English.

Briefly, Lefebvre proposes that space is produced through three interconnected processes: spatial practice, representations of space and representational spaces. With this proposition Lefebvre establishes the basis for 'trialectical' thinking; a way of conceiving space that includes not only its abstract qualities (measurements, coordinates, etc.) but also its historical and social dimensions. That is why Soja welcomes the idea of the Third Space and adapts it as one of the main theoretical underpinnings of his work – in fact *Thirdspace* is the title of one of his books. For Soja,

> the concept of the Third Space provides a different kind of thinking about the meaning and significance of space and those related concepts that compose and comprise the inherent spatiality of human life: place location, locality, landscape, environment, home, city, region, territory and geography (Soja 2009: 50 and 1996: 1).

Soja also invokes Bhabha's notion of the Third Space – as well as other postcolonial theorists such as Gloria Azaldúa, Edward Said and Gayatri Spivak – in order to address issues regarding minority and migrant groups, their cultural productivity and their modes of appropriating and creating space in contemporary cities. However, Soja does not develop a critique of Bhabha's work, or that of any of the postcolonial critics listed above, which would be necessary, at least, to explain how he adapts their ideas into his own work. Soja quotes extensively from all of them but fails to produce his own interpretation. Instead, Soja teases his readers with enthusiastic affirmations such as that the Third Space is 'a space where issues of race, class and gender can be addressed simultaneously without privileging one over the other' (Soja 2009: 50). Although Soja has a clear affinity with Bhabha in this respect, Soja does not move forward the discussion. He describes the work of a select group of minority artists and theorists in the United States, implying that it demonstrates the emergence of alternative minority movements, such as Chicanismo, which presumably are examples of Third Spaces. Soja rightly suggests that the concept of the Third Space provides an opportunity to address issues regarding minority peoples but he seems only able to do so via the work of artists and other critics who, as such, are already somewhat detached from the ethnic and cultural minorities that they represent.

In order to overcome Soja's inability to account for the cultural productivity of minority peoples, it is necessary to change the focus of attention. Rather than concentrating on the work of artists and theorists who occupy a privileged position in the hierarchical structures of trans-national cultural flow, consideration should be given to the products of lay people who live, physically and metaphorically, on the periphery or invisibly in the nooks and crannies of contemporary cultures and cities. There is no better or more accurate place to locate those minority peoples than in the slums of contemporary cities around the world; those areas where the figure of the Third Space materialises itself in seemingly endless neighbourhoods and squatter settlements where cultural meaning is most certainly reconstituted constantly. These are spaces where the certainties and norms of the contemporary world-order are bluntly rendered inadequate. Indeed, the Third Spaces where the poor live transcend the dualism of east and west, periphery and centre, third and first world. I refer here to extensive areas of abandoned and derelict housing which are used by squatters, gangsters and youngsters in cities such as Liverpool in the United Kingdom, or the dramatic consolidation of supposedly temporary settlements in post-Katrina New Orleans. I also refer here to the immense peripheral settlements where more than half the population of cities in Africa, Latin America and India live.

These are spaces where the certainties and norms of the contemporary world-order are bluntly rendered inadequate.

These are spaces of encounter where peoples from different origins meet, where different economies develop, where our understanding of 'city', its meaning, proves insufficient to subsume the proliferation of antagonistic lifestyles and spatialities. Needless to say, these are spaces where great imbalances in the distribution of power become visible – not in the past but today. Their position in relation to the formal, dominant or hegemonic city is ambiguous; they are part of cities but they are also excluded from them (for example, in Brasilia, Caracas or Nairobi). Yet, in all cases, slums are examples of the creativity of the people, their ability to produce new forms of knowledge and cultural expression in order to cope with the demands of urban living and a globalised market economy that has no place for the poor. It is in this context where the concept of the Third Space could contribute to the development of architectural theory. Bhabha's theorisation of the Third Space provides a possibility to inscribe, and to validate architecturally, the actions of the people as producers of their own inhabitable space, as true agents in the continued reshaping of cities around the world.

CHAPTER 6

The Pedagogical and the Performative

Unlike hybridity, a term that has been used frequently in architectural history and theory, the terms pedagogical and performative, which Bhabha uses in his critique of the nation, have not had a great deal of impact on contemporary architectural debates. However, much like hybridity and the Third Space, the pedagogical and the performative offer ample opportunity to address some crucial aspects which have received comparatively less attention in architectural studies. Amongst such aspects is the way in which colonial cities have been historicised and, more recently, the way in which architectural informality in cities throughout the world (slums, squatter settlements and so on) has – or, rather, has not – been studied by architects and historians. These are only two examples of the many thematic areas where Bhabha's discussion of the pedagogical and the performative may help to advance discussions on contemporary architecture.

Before explaining Bhabha's critique of the nation and the implications of his chosen terminology, it is important to elaborate on two concepts: cultural difference and the idea of nation itself. Cultural difference is a strategic notion in Bhabha's writing. I referred to it briefly in the introduction, and have used the term repeatedly throughout the book but, so far, I have not offered an appropriate explanation of what it means. I have located the explanation of this term at the end because it is a powerful vehicle to make the transition from a historical focus on colonialism (as in the previous chapters) to the analysis of contemporary cultures or, else, the particularities of cultural interaction today. Besides, cultural difference is a key concept in Bhabha's critique of the nation, as well as in the theoretical empowering of people as its constituents. Following the explanation of cultural difference I will proceed to review the idea of nation by drawing on the work of historians who have influenced Bhabha: Eric Hobsbawm and Benedict Anderson. Their work helps to develop our understanding of the nation as a political construct and explains the reason why Bhabha criticises the concept of nation so acutely. After explaining the concept of cultural difference and reviewing the idea of the modern nation, I will return to analyse Bhabha's critique of the nation, and the implications of the terms pedagogical and performative in that critique.

The irredeemable conflict that Bhabha illustrates via the terms pedagogical and performative provides fertile ground for the study of contemporary architectures, both in previously colonised countries as well as elsewhere. His mode of questioning the homogeneity of the nation, and its reliance on history as a linear fabrication, offers an opportunity to scrutinise the narrative construction of architectural 'historicity' and the idea of national identity in architecture. The final part of this chapter shows how Bhabha's critique can contribute to the development of architectural history, theory and professional practice. I will analyse three specific case studies in each of the above-mentioned areas: history, theory and professional practice respectively, and in three parts of the world. With this analysis I wish to demonstrate that Bhabha's critique of the principles that underpin the concept of the modern nation can be appropriated in order to advance an interrogation of architecture as a pedagogical narrative created on the basis of a linear history that joins contemporary architectural production around the world with Europe's past – Greek, Roman, Medieval, Renaissance, Baroque, Modern architectures and so on.

Cultural difference and the agency of minorities

Cultural difference is one of Bhabha's most potent and politically charged concepts. Bhabha introduces the concept of 'cultural difference' in opposition to 'cultural diversity', a term that he finds not only misleading but also derogatory. In Bhabha's view, the idea of cultural diversity belongs to the western liberal tradition which places individual freedom, as well as the rights and equality of

Cultural difference is one of Bhabha's most potent and politically charged concepts.

mankind, at the basis of society. Thus, western liberalism not only accepts that cultures are diverse but, also, promotes the coexistence of diverse cultures as part of the development of national societies. However, Bhabha argues that while cultural diversity is encouraged in most liberal democratic societies, there is also a correspondent containment of true diversity. Containment materialises itself in two forms: it either conditions the presence of other cultures within the space of the nation, or it totalises cultures. The first form of containment refers to the fact that the presence of other cultures within the national territory is permitted only as long as those 'others' conform to the parameters established by the host society for their interaction. As Bhabha himself puts it, 'those other cultures are fine, but we [the host society] must be able to locate them within our grid' (Bhabha 1990: 208). In practical terms, containment is achieved, amongst other ways, through issuing visas and work permits, by establishing language requirements and nationality tests or by creating integration and assimilation committees. These procedures allow the location and classification of 'difference' within a system of multicultural coexistence determined by dominant societies. Therefore, cultural diversity belongs – although it pretends otherwise – to a rhetoric of cultural stratification whose aim is to enforce conformity to an idealised version of the nation as homogeneous, a grid. The 'grid' is an ingeniously chosen term because it is a rational system. As such, it presupposes the rationalisation of cultures so that they can be inscribed into another system (the host culture) that has also been abstracted in order to permit their mutual interaction. That is why, Bhabha argues, 'cultural diversity is an epistemological object – culture as an object of empirical knowledge' (Bhabha 1994: 34). This affirmation brings us to the second form of containment, the totalisation of cultures, because for cultural diversity to work in the manner previously explained cultures need to be conceived as complete homogeneous entities: *the* Chinese, *the* Indian, *the* Spanish.

This conception of cultural diversity results in what Bhabha provocatively calls the *musée imaginaire*, where cultures can be appreciated, appropriated, located geographically and historically, as if they were in a museum. In this context, the term *musée imaginaire*, which translated into English means imaginary museum, or museum of the imagination, refers to a nineteenth-century (and early twentieth-century) debate about the exhibition of foreign objects (mostly from the colonies) in European museums. Many of these objects were not created for aesthetic contemplation but for daily use, they were utensils, tools or religious figures. Consequently, the exhibition of foreign, 'primitive' objects in the museum amounts to the inscription of other cultures into the western 'grid'. In so doing, the west normalises its own history of colonial expansion and exploitation by inscribing the history of colonised subjects into its own. With this argument Bhabha maintains that cultural diversity is conducive to, and facilitates, the commodification of cultures in a global market place – which is another form of containment, because only the wealthy can afford them.

Bhabha goes on to discuss an issue that is very relevant today: racism. He stresses that 'in societies where multiculturalism is encouraged racism is still rampant in various forms' (Bhabha 1990: 208). Though it is paradoxical, racism and xenophobia result because the presence of other people within the space of the nation comes to be seen as a menace to the people of the host society who fear the loss of their 'national' identity as well as the disruption of their welfare and, even, national security. In turn, menace transforms itself into aggression against the ethnically different (racism), or the foreign in general (xenophobia). That is why Bhabha concludes that 'cultural diversity is also the representation of a radical rhetoric of the separation of totalised cultures that live unsullied by the intertextuality of their historical locations, safe in the utopianism of a mythic memory of a unique collective identity' (Bhabha 1994: 34). If this statement needs explanation it is only to clarify two concepts: 'unsullied by the intertextuality' and 'the utopianism of a mythic memory'. The former refers to culture as epistemological objects (totalised) and separate, so that they remain free from contamination (unsullied) by interaction with other cultures (intertextuality); so that *the* Chinenese, *the* Indian or *the* Spanish remain pure and unique. The latter focuses on culture as an idealised concept whose purity and homogeneity can only be found in versions of history that omit or deny difference, as in Macaulay's representation of *the* British (Chapter 4).

In contrast, Bhabha's notion of cultural difference does not attempt to totalise cultures in order to rationalise their interaction within a particular geo-political

... Bhabha's notion of cultural difference does not attempt to totalise cultures in order to rationalise their interaction within a particular geo-political frame (or grid).

frame (or grid). Quite the opposite, cultural difference reveals the fact that difference is an inherent part of all cultures, that no culture is homogeneous but heterogeneous. Unlike cultural diversity and multiculturalism, cultural difference does not refer to the 'free play of polarities and pluralities in the homogenous empty time of national community' (Bhabha 1994: 162). Instead, Bhabha maintains, 'the question of cultural difference faces us with a disposition of knowledges or a distribution of practices that exist beside each other, ... designating a form of social contradiction or antagonism that has to be negotiated rather than sublated' (Bhabha 1994: 192). Thus, the concept of cultural difference proposes a strategy for the articulation and continued negotiation of contesting cultural sites. Rather than a totalised object, Bhabha turns culture into the perennial process of its own construction. To put it simply, because cultures are formed by, and emerge from, the continual articulation of antagonistic positions, it is possible to affirm that they are also always incomplete – the reason why Bhabha wants us to understand culture as a process rather than as a pre-given content or a sign of certainty. *The* Chinese, *the* Indian, *the* Spanish, can no longer be understood in the finitude of the banner but through the conflictive histories of the different 'peoples' those banners represent. That is why Bhabha suggests that the notion of cultural difference opens up a space of cultural contestation: because it gives presence to minority positions that remain subjugated by official conceptualisations of the nation. Minority positions include not only ethnic groups (usually foreigners) but, also, women, economic migrants, exiles, gays and lesbians, and the poor. Indeed the poor are often ignored in debates about minorities – not to mention architectural debates in which the poor do occupy a very unstable position. One reason for such a slippage is that the poor transcend boundaries across different minorities and, even, penetrate into the dominant culture. In other words, the poor are not just migrants, or ethnically different people, or women, gays or lesbians; the poor are also white, western-born individuals and, often, Christian. The fluidity of 'the poor' as sociological category disturbs, more than any other minority, the cohesion of the nation as a rational socio-political system of collective control. As such, the poor are a prime example of the existence of cultural difference and its effect in the continued construction of all aspects of the nation: culture, politics, race, etc. It may also be possible to understand at this point why cultural difference is one of Bhabha's most potent and politically charged concepts and, also, why it lies at the centre of his critique of the modern nation.

The nation and some ideas on nationalism

There is a tendency to believe that nations, as we know them today, have existed for centuries. However, historically, this is not so. The nation, as a socio-political and administrative construct, is relatively new in historical terms. In fact, they are a product of the Enlightenment and largely also of colonialism. The origin of the nation-state can be traced to the demise of classical and medieval empires as well as to the obsolescence of the city-state form of administration. This locates the historical emergence of the nation-state in the late Renaissance, a period referred to by contemporary scholars as 'early modern'. However, Eric Hobsbawm and Benedict Anderson observe that nationalism – and nationalist ideologies and movements – only gained momentum in the nineteenth century, the period when most existing European nations were formed. In his book *The Age of Empire 1875–1914*, Hobsbawm offers a compelling account of the development of nations and nationalism during the period in question (1875–1914). The two terms are inseparable from each other because the nation results from the aspirations of nationalist movements whose interest is to represent a community of people who inhabit a particular territory, share the same descent (race and history) and speak the same language. This is the ethno-linguistic definition of the nation we are familiar with. However, Hobsbawm

The nation, as a socio-political and administrative construct, is relatively new in historical terms.

explains that until the beginning of the eighteenth century, when most societies were agrarian, people had a different relation with the territory they inhabited. Farming, and all the activities intrinsically linked with it, was the basis of what Hobsbawm calls '*real* communities of human beings with *real* social relations with each other' (Hobsbawm 1987: 148). Towards the end of the century rapid industrialisation caused significant changes in the methods of agrarian production and consumption. Additional pressures arose from the consolidation of international trade networks with the colonies and other European empires. These contingencies required different methods of administration. The porous and imprecise territoriality of the Middle Ages, with people who professed different allegiances (political, religious, linguistic, etc.), was no longer sufficient, and so the feudal system receded. With the decline of agrarian communities, Hobsbawm asserts, 'nationalism and the state took over the associations of kin, neighbours and home ground [and replaced them] for territories and populations of a size and scale which turned them into metaphors' (Hobsbawm 1987: 148). By metaphor, a term that Bhabha appropriates readily in his discussion about the nation, Hobsbawm refers to the nation as a figure that transfers the meaning of community to the nation-state as a political construct. That is why, for Hobsbawm, the nation is an imaginary community which fills the void left by the decline of 'real' communities. The nation is imaginary because it 'creates some sort of bond between members of a population of tens – today even hundreds – of millions' (Hobsbawm 1987: 148) who neither know each other, nor have heard of each other, and never will.

The nation is imaginary because it 'creates some sort of bond between members of a population of tens – today even hundreds – of millions' (Hobsbawm 1987: 148) who neither know each other, nor have heard of each other, and never will.

In order actually to construct the nation, it is necessary to educate people: to instruct them on how to be good subjects and citizens. Education became an essential means for the advance of nationalism and the consolidation of the nation-state. In turn, national education required a language of instruction and, so, 'education joined the law courts and bureaucracy as a force which made language into the primary condition of nationality' (Hobsbawm 1987: 150). Education, then, turned into a means of linguistic and cultural homogenisation of citizens for the purpose of national administration. At the same time, homogenisation led to the exclusion of certain groups or individuals: those who do not wish to comply and those who 'cannot' be granted membership. 'In short', Hobsbawm says, the nation-state 'helped to define the nationalities excluded from the official nationality by separating out those communities which, for whatever reason, resisted the official public language and ideology' (Hobsbawm 1987: 150–1). This is an indication that the ambivalent strategies used to construct the nation-state correspond with the process of inclusion and exclusion that Bhabha identifies in the construction of colonial authority, and the reason why he decides to include the modern nation in his critique (see Chapter 4). Like Bhabha, Hobsbawm also discusses the contradiction of educating colonised subjects in the language and culture of their masters for the purpose of administration (private and public) while denying them equal treatment. 'Colonial peoples were an extreme case', says Hobsbawm:

> Since it was clear from the outset that, given the pervasive racism of bourgeois society, no amount of assimilation would turn men with dark skins into 'real' Englishmen, Belgians or Dutchmen, even if they had as much money and noble blood and as much taste for sports as the European nobility – as was the case with many an Indian rajah educated in Britain (Hobsbawm 1987: 152).

Racial discrimination ratifies the contradiction inherent in nationalist discourses which aim to form homogeneous societies by means of language and ethnicity. Unlike real communities, as Hobsbawm calls them, the modern nation-state is considered an artificial construct created mostly for the purpose of facilitating governance and promoting economic development. For that reason, nations are also culturally unstable and, in fact, ambivalent organisations: they congregate, under the myth of nationhood, diverse groups of people who do not fully correspond with the homogenising signifier of 'the nation' (ethnic groups and social classes, for example, as well as clans, religions, genders and so on).

Although Hobsbawm describes the nation as an imagined community, the term is more commonly linked to Benedict Anderson, author of the book *Imagined Communities: Reflections of the Origins and Spread of Nationalism*. In his book Anderson advances the definition of the nation 'as an imagined political community' (Anderson 2006: 6). Unlike Hobsbawm, who uses the term 'imagined' somewhat tentatively – that is, in opposition to real communities but also in order to highlight the difficulties of defining the term 'nation' itself – Anderson's usage is more precise. For Anderson the nation is not imagined in opposition to other 'real' or 'true' kinds of communities. The nation is always an imagined community and there are different imaginings, or forms of imagining it. In his view, there is no other way of conceiving the nation than as an artificial, abstract and, therefore, imagined form of social coexistence; which is not to say that the nation does not materialise itself in various forms, for example, through national constitutions, national legislation, nationality documents, monuments, books, maps and, more solidly, walls and fences between nations. There is, therefore, a disparity between the nation as imagined and the way it appears before us: one is abstract (intangible), the other is concrete and exclusive. The contradictions inherent in the construction of the nation (the inclusion/exclusion of people and the abstract, all-inclusive concept of nation versus the certitude of its concrete manifestation) motivate Bhabha's critique, which is not a critique of nationalism but of the way in which people are inscribed in the narratives of both nation and nationalism. Indeed, the fact that Anderson advances his definition of the nation as an imagined community in 'an anthropological spirit' fits Bhabha's critique well, because it is the people of the community who

… for Bhabha, the word 'imagined' erases the conflicts and complex historical experiences of the people who make the national community.

contradict the apparent cohesion of its imagining. In other words, for Bhabha, the word 'imagined' erases the conflicts and complex historical experiences of the people who make the national community.

Bhabha's critique of the nation

In his essay 'DissemiNation: Time, Narrative and the Margins of the Modern Nation', Bhabha advances a critique of the concept of nation as an artificial construct, a narrative, which eliminates cultural difference in its attempt to represent people as a homogeneous body. The essay was published originally in Bhabha's edited volume *Nation and Narration* (1990) and, later, in his book *The Location of Culture* (1994). The title of the former book indicates quite clearly the way in which Bhabha tackles theoretically the question about the nation as a narrative structure which assists in the construction of national identity; a narrative that makes national identity intelligible to us all. In fact, he often talks about the 'act of writing the nation', because it is primarily through writing that the nation emerges as a cultural construction. The Argentinian epic poem *Martin Fierro*, by José Hernández, comes quickly to mind as an example. The poem tells the story of Martin Fierro, an impoverished Gaucho (the inhabitants of the Argentine Pampa) who deserts the military and lives a troubled existence persecuted by the militia and distrusted by the natives. Thus, the protagonist lives on the fringe between the cultures, laws and traditions of both the Europeans and the natives. Indeed Gauchos themselves are people of mixed origin; either criollo (Spanish and Portuguese born in the Americas) or mestizo (the result of inter-breeding between Europeans and indigenes). As such, the Gauchos represent the reality of modern Argentinian culture, the fact that it is different from the European but not quite indigenous either. What is more, the Gauchos played an important role during the wars of Argentinian independence and, so, are pivotal figures in the national history. Thus, *Martin Fierro* is a narrative which asserts the contribution of the Gauchos to the development of Argentina and its collective contemporary national identity. Another example, perhaps better known, is the film *Braveheart* (1995) which tells the story of the legendary Scottish hero William Wallace who fought against the English during the First War of Scottish Independence. Even though the film was produced recently, it clearly evokes nationalist sentiments by appealing to the courage of the Scottish people while asserting their ethnicity (and some aspects of their language). Another nationalist narrative is found in Delacroix's painting *Liberty Leading the People* (1830) which portrays a bare-breasted Marianne (already a symbol of France) as a goddess-like figure holding the French flag with her right arm and a bayoneted musket on the other. Marianne, who stands dominant in the centre of the canvas, is being followed by a crowd of men of different ages and social classes, all of whom are fighting together for freedom. Delacroix's painting embodies the spirit of the revolution and its motto: *liberté, égalité, fraternité*. All these narratives which recount historical events, with different degrees of accuracy, help to create and to reinforce a sense of national identity and belonging. They contribute to the narrative construction of the nation by asserting racial (ethnic), linguistic and geographic features of the nation, features which unite the people even though they do not know each other.

... people are not *representative* of the nation but *represented* by national narratives in ways that are beneficial for the cohesion, prosperity and governance of the nation.

The discourse of nationalism as such is not Bhabha's main concern. The purpose of his essay is to formulate 'the complex strategies of cultural identification and discursive address that function in the name of "the people" and "the nation" and make them the immanent subjects of a range of social and literary narratives' (Bhabha 1994: 140). The expression 'function *in the name* of the people and the nation' contains an important clue to understanding the focus of Bhabha's critique. It suggests that strategies of cultural identification and discursive address are external to both people and nation. They act on behalf of the people and the nation making them the ever-present subjects of social and literary narratives (about the nation). Consequently, it is not the people who represent themselves as constituents of the nation, but the narratives which represent both the people and the nation they constitute. In other words, people are not *representative* of the nation but *represented* by national narratives in ways that are beneficial for the cohesion, prosperity and governance of the nation. A second clue lies in the way Bhabha singularises 'the people' and 'the nation'. In doing so, Bhabha stresses his discomfort with totalisation, or the holistic representation of bodies which are inherently heterogeneous: people and nation.

An aspect that causes Bhabha a great deal of theoretical disquiet is the reliance on history or, as he puts it, 'the historical certainty' evoked by the concept of nation and the narratives that sustain it:

> My emphasis on the temporal dimension in the inscription of these political entities ['the people' and 'the nation'] – that are also potent symbolic and effective sources of cultural identity – serves to displace the historicism that has dominated discussions of the nation as a cultural force. The linear equivalence of event and idea that historicism proposes, most commonly signifies a people, a nation, or a national culture as an empirical sociological category or holistic cultural entity (Bhabha 1994: 140).

Rather than history as such, Bhabha refers to historicism, that is, the notion that history is an endless succession of cumulative events according to which the present of any society, and all the social activities of that society (art, science, philosophy, architecture, etc.), is defined by their history. This view of historicism is closely linked with the work of the German philosopher Georg W. F. Hegel who interpreted the progress of mankind as a historically cyclical process, a linear progression of causes and reactions – hence the name 'dialectic' given to his philosophical method. Bhabha's reference to the equivalence between event and idea is, undoubtedly, a reference to Hegelian dialectic historicism and a critique of it. For Bhabha, such an interpretation of history reduces the nation, its society (the people) and its culture to empirical categories or totalities (as explained above). He argues that claims for the homogeneity of the nation are only tenable on the basis of such a kind of reductive historicism. Suitable examples are Macaulay's representation of the English as a culture whose origin can be traced to Ancient Greece and, so, hides all its internal differences (see Chapters 2 and 4). Macaulay's example shows how a narrative correspondence is made between the past and the desired imagining of the nation's present. Bhabha's main argument is that such a narrative correspondence is often artificial, a historical fabrication which serves the purpose of eliminating difference in order to grant the socio-political solidity of the nation. Moreover, from a postcolonial perspective, such a mode of historicisation (linear, homogeneous and serene) obscures, even denies, the violent realities of the nation's colonial origins; a fact which needs to be uncovered, because the nation, as a form of political administration, is inseparable from colonialism, that is, from the contingencies of international trade, governance and the need of establishing modes of cultural supremacy to govern the peoples of foreign territories, the European subjects.

Adhering to Benedict Anderson, who observes that there is an inherent contradiction between the relative historical newness of the nation (both as a concept and as a political entity) and the antiquity that nationalist discourses assigned to it, Bhabha asks 'how does one write the nation's modernity as the event of the everyday and the advent of the epochal?' (Bhabha 1994: 141). Bhabha's question addresses the contradiction of conceiving the nation through the incongruous set of contemporary experiences that we have of it, while simultaneously contemplating the certainty conveyed by the nation's historicity. In plain terms, the question refers to the paradox of walking along Brick Lane, in East London, where there are street signs in various languages; where products from multiple countries are on display; where people of countless racial backgrounds speak an unthinkable amount of languages while eating 'chicken tikka masala' with French wine; and yet one is asked to be certain of being in the same England which, according to Macaulay, descends unequivocally from classical Greece.

Bhabha's question addresses the contradiction of conceiving the nation through the incongruous set of contemporary experiences that we have of it, while simultaneously contemplating the certainty conveyed by the nation's historicity.

The contradiction between the newness of the nation and its alleged antiquity or, in Bhabha's terms, between the nation as the event of the everyday and advent of the epochal, is what he calls the 'ambivalent temporality of the nation-space': the simultaneity of two national times which appear to contradict each other. 'Such an apprehension of the "double and split" time of national representation', Bhabha says, 'leads us to question the homogenous and horizontal view associated with the nation's imagined community' (Bhabha 1994: 144). Here homogeneity refers to the similarity amongst the members of the community (racial, linguistic, religious similarity) while horizontality suggest that there are no hierarchical differences amongst the people of the nation. Since that is clearly not the case, Bhabha argues that nationalist narratives of holism and social cohesion obscure the realities and fragmented histories of the nation's people, who are conceived and inscribed in 'double-time':

> We then have a contested conceptual territory where the nation's people must be thought in double-time; the people are the historical 'objects' of a nationalist pedagogy, giving the discourse an authority that is based on the pre-given or constituted historical origin in *the past*; the people are also the 'subjects' of a process of signification that must erase any prior or originary presence of the nation-people to demonstrate the prodigious, living principles of the people as contemporaneity: a sign of the present through which national life is redeemed and iterated as a reproductive process (Bhabha 1994: 145).

With this statement Bhabha shifts the focus to the people as the constituents, or signifiers, of the nation. The contradiction that we just addressed above becomes apparent in the fact that the people are inscribed both as the object of historical national narratives and the subjects of the nation's current (multi)cultural vitality. The Gaucho mestizo who lives in the vast Argentine Pampa, or the white, brave, Gaelic-speaking Scottish, or the revolutionary and egalitarian French are juxtaposed with images of 'multiculturalism' which demonstrate the vitality, prosperity and liberalism of the modern nation in a globalised economy. In these cases people are the historical objects of a national discourse whose authority is constituted in the past and, also, the subjects of free, culturally vital and forward-looking nations in the present. The homogeneity inherent in the first does not correspond with the plurality and dynamism experienced in the second. Hence, Bhabha argues, people have to be conceived in double-time.

To elucidate further this argument Bhabha puts forward the two temporalities of the nation: the pedagogical and the performative. The reason why the pedagogical and the performative are called temporalities rather than dimensions or layers is because they refer to the above-mentioned different moments: the past and the present. The pedagogical refers to the nation's history (the pre-given, the past) while the performative refers to the 'people as contemporaneity', a sign of the present. Bhabha maintains that the double inscription of the people – as signifiers of both past and present – in the production of national narratives creates 'a split between the continuist, accumulative temporality of the pedagogical, and the repetitious, recursive strategy of the performative' (Bhabha 1994: 145). Let us explain. The pedagogical temporality corresponds with the official project of the nation as historicity and self-generation; that is, the linear chronological histories that are taught in schools, in the national languages of every country, along with the geographical understanding of the nation's boundaries, its laws, its heroes, its traditions and so on – the nation as represented in the textbook or even fictionally, as in the three previous examples (*Martin Fierro*, *Braveheart* and *Liberty Leading the People*). In short, the pedagogical refers to all those instruments used to inculcate a sense of national identity and belonging, the means through which we recognise ourselves as members of an imagined community. The performative temporality, on the other hand, brings to the fore the people as agents of a process of national signification which renders the homogenising intent of the nation's narrative both inappropriate and unrealisable. As such, the performative temporality can be understood as the

> The pedagogical refers to the nation's history (the pre-given, the past) while the performative refers to the 'people as contemporaneity', a sign of the present.

anti-official or, as Bhabha puts it, 'a counter-narrative of the nation that continually evokes and erases its totalising boundaries – both actual and conceptual – disturbs those ideological *manoeuvres* through which "imagined communities" are given essentialist identities' (Bhabha 1994: 149). To put it simply, as the term indicates, the performative refers to the entire set of actions (artistic, commercial, political, religious, etc.) carried out by people in their everyday life. The performative can be understood as the staging (Bhabha uses the French expression *mise en scène*) of cultural difference. The effect of the performative is anti-official, or antagonistic, because the political unity of the nation resides in the permanent negation of its plurality (its cultural difference); or, as Bhabha puts it, in 'the continual displacement of the anxiety of its irredeemably plural modern space' (Bhabha 1994: 149). As mentioned above, by means of the performative temporality, Bhabha turns his attention to the peoples of the nation and, so, he opens up pockets, or areas, within the nation-space which allow cultural difference to emerge. In other words, the performative temporality draws attention to those 'recesses of national culture' which are obscured by the nation's pedagogical narrative: women, ethnic minorities, youth culture, social movements, etc. These minority groups, Bhabha asserts, 'assign new meanings and different directions to the processes of historical change' (Bhabha 1990a: 3).

We are now returning to the concept of cultural difference with which this chapter began. We said that cultural difference reveals that no culture is a single culture but, instead, that all are multiple and plural. Hence, the Other is no longer found outside the nation, in other nations, but in its own internal heterogeneity:

> The problem is not simply the 'self-hood' of the nation as opposed to the otherness of other nations. We are confronted with the nation split within itself, articulating the heterogeneity of its own population. The barred Nation It/Self, alienated from its eternal self-generation, becomes a liminal space that is internally marked by the discourses of minorities, the heterogeneous histories of contending peoples, the antagonistic authorities and the tense locations of cultural difference (Bhabha 1994: 148).

The problem is not simply the 'self-hood' of the nation as opposed to the otherness of other nations. We are confronted with the nation split within itself, articulating the heterogeneity of its own population.

Once again, Bhabha dismantles the primacy of binary systems according to which the 'Other' of the nation is external to it (i.e. the Other of England is France). Instead, Bhabha proposes that difference (or otherness) is an inherent part of national culture. However, it is important not to understand Bhabha's proposal as a replacement of the otherness outside for its otherness within and, so, to consider minorities as an undifferentiated mass of people inside the national territory. On the contrary, by foregrounding the nation's internal difference (its otherness within) Bhabha asks us to recognise that nations are inhabited by multiple groups which not only perform their differences but also interact constantly and renegotiate the boundaries that separate them. Their presence within the nation disturbs its conceived (pedagogical) homogeneity and prevents the full accomplishment of nationalist discourses: because they make visible that the nation is marked by the presence contesting cultural positions-cultural difference.

Bhabha is once again at pains to open up spaces of contestation from where to challenge the homogenising intent of nationalist narratives. His theoretical objective is to authorise minorities as constituencies of the nation, to give them voice in the enunciation of the nation's narrative and in the permanent production of its culture; which is always cultures, plural. Consequently, minority groups, as well as the dynamics of interaction amongst them, contest – actively and by implication – the solidity of the nation and the national culture. In his own words:

> Once the liminality of the nation-space is established, and its signifying difference is turned from the boundary 'outside' to its finitude 'within', the

threat of cultural difference is no longer a problem of 'other' people. The national subject splits in the ethnographic perspective of culture's contemporaneity and provides both a theoretical position and a narrative authority for marginal voices or minority discourse. They no longer need to address their strategies of opposition to a horizon of 'hegemony' that is envisaged as horizontal and homogenous (Bhabha 1994: 150).

This is not an optimistic statement about the resolution of cultural difference and the elimination of hierarchical and social antagonism. The fact that Bhabha speaks of the nation as a liminal figure which prevents claims for transcendental authority, or hegemony, does not mean that the contest for authority is brought to a halt. On the contrary, the never-ending conflict between the pedagogical and the performative is a sign of the continued contest for narrative authority in the construction of the nation. The concept of cultural difference suggests a 'politics of difference' which consists of a permanent state of negotiation, not elimination, as the most productive way to tackle situations of social, cultural and political antagonism – negotiation, however, does not always imply resolution (see Chapter 5).

It is also important to make clear that Bhabha is not interested in undermining the nation as a political entity. He is aware that the current conditions of economic globalisation give validity to the nation as the most powerful form of social and political organisation. Nevertheless, Bhabha prompts a revision of the nation from the perspectives of minority groups taking also into consideration the historical experiences of previously colonised peoples because, for him, the modern nation cannot be separated from its colonial past. In reality, colonialism lies 'hidden' at the basis of the uneven distribution of power and wealth in the contemporary world.

Bhabha prompts a revision of the nation from the perspectives of minority groups taking also into consideration the historical experiences of previously colonised peoples ...

Questioning duality in the history of colonial cities

Bhabha's discussion about the pedagogical and the performative as two contesting but constitutive temporalities of the modern nation emerges as a useful model to question the methods that have been used to historicise cities whose morphological development has been determined by colonialism. As mentioned in Chapters 3 and 4, non-western architectures have often been studied – and inscribed historically – on the basis of a binary system of antagonism between the pre-colonial and the European or, more recently, between Euro-American modern and non-western architectures. The dismantling of binary systems and the foregrounding of cultural difference that Bhabha advocates brings about an opportunity to overcome such antagonism and, more importantly, to examine the contribution of other groups whose participation in the constant reshaping of cities has been ignored.

… some architectural critics argue along with Bhabha that binary methods of analysis are insufficient fully to address the complex urban conditions of cities whose formal development was severely determined by colonialism.

Indeed, some architectural critics argue along with Bhabha that binary methods of analysis are insufficient fully to address the complex urban conditions of cities whose formal development was severely determined by colonialism. Amongst such scholars is Brenda S. A. Yeoh who, in her book *Contesting Space: Power Relations in the Urban Built Environment in Colonial Singapore*, criticises acutely the dual approach to the study of colonial cities. Yeoh contends that representing the morphology of colonial cities as binary is both inadequate and distorting because it ignores the force and impact of colonialism (the very violence of the process of colonisation). She goes on to assert that the morphological features of cities which were drastically altered by colonialism cannot be understood separately from their central function in establishing, systematising and maintaining colonial rule. Here

Yeoh refers to the fact that the ordering principles of colonial cities, as designed and built by the coloniser, correspond with the creation of a hierarchical society in which the Europeans rule. In other words, the geometrical layout of cities is analogous to the desired social order the planners want to impose – or, as explained in previous chapters, the desire to see themselves repeated in the colonised (narcissistic demand). That is why the colonial city, both the concept and its realisation, needs to be understood as an affirmation of colonial power – an iteration of colonial discourse. The rational 'morphology' of the colonial city serves the purpose of diminishing the architectural values of 'amorphous' pre-colonial settlements, which are considered a sign of backwardness.

After establishing that the morphological features of the colonial cities cannot be isolated from their function as centres of colonial rule, Yeoh proposes that there are three features which distinguish colonial cities. The first is their extraordinary pluralism, colonial cities contain 'a diversity of peoples, including colonialists, immigrants, and indigenes intermeshed within a social matrix comprising newly constituted relations of domination and dependence between individuals and between collectivities of people' (Yeoh 1996: 1). We can add to Yeoh's list slaves, merchants, prostitutes and travellers, all of whom participated actively in the consolidation of colonial cities and their economy; that is, their form and socio-political function. In order to qualify the previous quotation, Yeoh affirms that 'these social groups are derived from vastly different societies, each with its own ingrained cultural behaviour, civil traditions and institutionalised practices' (Yeoh 1996: 1). Not only does Yeoh recognise the existence of different groups, more importantly, she also authorises their contribution to the formation of colonial cities. With this Yeoh establishes categorically that the history of cities whose development was affected by colonialism cannot be written in terms of a binary system of social or morphological development.

The second feature is that the system of social classification in colonial cities resembles neither the class structure of the European, nor the pre-colonial systems of social stratification of colonised peoples. With this Yeoh subscribes to the position that colonialism generates changes in the socio-political and cultural structures of all groups involved. None of these groups can return to its condition before colonisation nor can anyone ever achieve their desire of becoming like the other; the coloniser cannot achieve its goal of repeating itself in the colonised, nor will the colonised ever achieved the same status of the coloniser, they all enter a process of hybridisation which exacerbates itself exponentially the more they interact. And, of course, as the economic dimension of colonialism gains momentum (i.e. trade, demand, consumption, capital flow, etc.), interaction between all the parties involved cannot be stopped; neither can the process of hybridisation. The third characteristic is the unbalanced distribution of power, concentrated largely in the hands of the coloniser, a situation which causes a perpetual struggle for authority.

Thus, for Yeoh, the city is a territory of contestation. Through her detailed analysis of historical documents, statistics, laws and planning regulations in Singapore, Yeoh demonstrates that the physical – as well as the socio-political and cultural – fabric of the city reflects the continuous power struggle of many actors (individual and collective) whose interventions have helped to shape cities, even when they are at the bottom of the power structure (i.e. slaves, indigenes, prostitutes and so on). By this means, Yeoh refutes strongly the idea that colonial cities, whether in South East Asia or elsewhere, are the creation of the European ruler alone. Yeoh affirms that:

> the built spaces of the colonial city were not simply shaped by dominant forces or powerful groups, but were continuously transformed by process of conflict and negotiation involving strategies and counter-strategies of colonial institutions of authority and different 'colonised' groups within the society.

Yeoh goes further to explain that:

> the built spaces of the colonial city were construed as sites of control and resistance, simultaneously drawn upon by, on the one hand, dominant groups to secure conceptual or instrumental control, and, on the other, subordinate groups to resist exclusionary definitions or tactics and to advance their own claims (Yeoh 1996: 313).

Thus, for Yeoh, the production of space in cities whose morphological development was marked by colonialism is part of a permanent process of divergence and negotiation between those who are in control and those who live in and use the city daily. Although Yeoh does not refer to Bhabha directly, she clearly employs a postcolonial method of critique in order to contest the univocality of European historical accounts according to which colonial cities are the result of a duality between European and non-European norms. In sum, Yeoh calls for the creation of more appropriate methods of historical inscription which account for all the parties involved in the development of cities in previously colonised nations. Indeed, for that reason, many of the questions that Yeoh raises are applicable to western cities, which are also plural, complex and historically layered.

The performative temporality of contemporary cities

Another theoretical model with striking similarities to Bhabha's conception of the two national temporalities can be found in the work of Rahul Mehrotra. Mehrotra is an architect and theorist who has worked extensively on questions about the development of cities in India. Unlike Yeoh, who advances a historical critique, Mehrotra focuses on the contemporary condition of cities in India and other parts of the developing world – although his work does not exclude entirely cases in the developed world. Drawing directly on the work of Bhabha, Mehrotra studies the apparent antagonism between the space of the city designed by architects and the space of the city as used by people. Thus, Mehrotra proposes that there are two different kinds of space, or two cities inextricably intertwined: the static and the kinetic. The static city is made of permanent materials such as concrete, brick, metal and wood. It is planned, or designed, according to pedagogical narratives (i.e. architectural discourse and urban

Drawing directly on the work of Bhabha, Mehrotra studies the apparent antagonism between the space of the city designed by architects and the space of the city as used by people.

legislation). Not only do such pedagogical narratives determine the city's image but, also, serve to make the city ordered, productive and governable. That is why Mehrotra maintains that 'architecture is the spectacle of the static city'. As a symbol of power and control (historically and in the present) the static city is conceived to be durable and stable. The kinetic city, on the other hand, refers to the performances of the people 'within' the physical confines of the static city. To illustrate his argument Mehrotra underlines the way in which 'processions, festivals, street vendors and dwellers, all result in an ever-transforming streetscape – a city in constant motion whose very physical fabric is characterised by its kinetic quality' (Mehrotra 2009: xi). To exemplify his point further, Mehrotra refers to events such as the Bazaars in the Victorian Arcades of Mumbai, where street vendors have appropriated a historic area of the city in order to create a spontaneous market which permits their economic survival. The physical appropriations of space carried out by the poor people who use the arcades and their performances (vendors, prostitutes, squatters, etc.) introduced great dynamism to the static city – represented, in this case, by the Victorian Arcades and the architecture of the historic Fort Area in Mumbai. The activities of the poor contradict the purpose of the arcades which were designed originally to protect pedestrians from the elements and to smooth their transit as they shopped in the stores on both sides of the road. Now that the arcades have been occupied by hawkers who compete with the 'proper' stores, pedestrians need to negotiate their way through a labyrinth of vendors, and often walk on the road, while being overwhelmed by people offering their wares, quoting prices, offering services, all in the midst of an enormous cultural exuberance. The kinetic, thus, re-signifies the static space in every aspect: architecturally, economically and socio-politically. As with Bhabha's concept of the performative, Mehrotra's notion of kinetic space brings forward the people, the multiple users who appropriate space creatively. Mehrotra points out that the Bazaar is a strategy of survival for the poor in an unstable Indian economy. On this basis, Mehrotra affirms that the kinetic city – rather than the static – is the most accurate representation of cities in the developing world. Indeed, Mehrotra's discussion about the static and the kinetic reflects Bhabha's interest in foregrounding the performances of the people, as well as their participation in the continued construction of national cultures and identities. Through their acts of appropriation, the inhabitants of Mumbai are considered the producers of social, cultural and physical spaces that represent the tense interaction between different groups and the conflictive socio-political and economic realities of India – and many other cities in the developing world. The concept of kinetic space is comparable with the performative because both serve to demonstrate that contemporary cities are internally marked by the discourses of the people, their contending and heterogeneous realities, all of which unavoidably affect the physical fabric of urban conglomerates around the world, not only in the non-west.

Architecture and the performative

So far, in the second part of this chapter, we have examined the possibilities brought forward by Bhabha's theory in order to challenge various issues regarding the historicisation and theorisation of architecture. In the case of Brenda Yeoh, we discussed the unsuitability of binary historicising methods according to which the morphology of former colonial cities is characterised by the superimposition of pre-colonial and European forms. Subsequently, we extended her historical analysis to contemporary cities, as in the case of Mumbai. Yeoh's and Mehrotra's theories show that cities have been shaped – physically as well as culturally, socially and politically – by a multiplicity of participants that includes Europeans and indigenes, as well as slaves, merchants, women, gays and lesbians, prostitutes and so on (I am trying to stress the fact that the terms minorities and cultural difference refer to categories other than ethnic). However, little has been said about the way in which this discussion could also challenge methods of architectural design. In order to address this complicated matter I will examine an interesting case study: the work of ELEMENTAL, a Chilean practice which specialises in social housing. Since the case is not well known internationally – though the practice is rapidly gaining worldwide recognition – I will commence with a brief description.

Such a position of adjacency makes it difficult to situate ELEMENTAL comfortably, and exclusively, as an architectural practice although they are known by the buildings they produce.

ELEMENTAL is a practice ambiguously situated on the margins between architectural research, practice and academia. Its principal, Alejandro Aravena, is an architect who teaches at the School of Architecture of the Universidad Catolica de Chile. Indeed, their office is located on the grounds of the school, even though the practice is a partnership between the University, COPEC

(Compañía de Petróleos de Chile) and the founders of ELEMENTAL. From this interesting position, ELEMENTAL carries out continued academic research on the living conditions of the urban poor in Chile, designs buildings to satisfy their needs and deals with local economic and governmental policies. Thus, ELEMENTAL fits quite precisely Bhabha's definition of interdisciplinarity: '[a] space for the articulation of cultural knowledges that are adjacent and adjunct but not necessarily accumulative, teleological or dialectic' (Bhabha 1994: 163). Such a position of adjacency makes it difficult to situate ELEMENTAL comfortably, and exclusively, as an architectural practice although they are known by the buildings they produce.

Working within a government framework created in 2001, ELEMENTAL does consultancy work for poor communities in cities throughout Chile and designs housing schemes for them. Their aim is to provide middle-income housing near city centres for people who are, statistically, incapable of repaying a mortgage. To do so, ELEMENTAL works against economic constraints – their maximum budget is US$10,000 per unit, not including the cost of land, an amount which allows only for the construction of approximately 30 square metres – and needs to transgress both legislation and public opinion. They challenge existing legislation because instead of relocating the poor to the peripheries of cities (where they do not disturb the desired image of cities' historic centres), they advocate city-centre living. They also transgress public opinion because, in keeping the poor near the city's core, they generally upset the middle and upper classes for whom the poor is an undifferentiated mass to be approached with great scepticism.

Since the budget is not sufficient to build a complete middle-class house, let alone purchase land near the city's core, jokingly (but realistically) Aravena says that they can only design one half of a house for each family. For ELEMENTAL, this situation does not limit the possibilities of design. On the contrary, they assert that it multiplies the possibilities: it presents a challenge for the architect and for architecture in general. To mitigate the lack of funding, the team led by Alejandro Aravena has devised a method of incremental urban development and densification which they call 'dynamic development in time'. ELEMENTAL designs a basic living unit, the rest will be developed by users without control by the architect. But, what half of the house do they build?

The answer varies depending on the circumstances of each project. The idea is to provide those services people would not be able to afford once their initial subsidies have run out and the architects (and other consultants) have gone. In most cases these services include the general layout of the area, a flexible housing structure and electrical and mechanical arrangements. With the general urban layout (the master plan, as it were) ELEMENTAL sets a few parameters for future development. For example, rows of houses are unusually distant from each other in order to create ample public spaces rather than simply streets. The purpose of such an arrangement is to facilitate the appropriation of space for different, often unexpected, activities. Rather than specifying particular uses, or creating spaces for clearly defined purposes, the master plans designed by ELEMENTAL arise from, and respond to, the wide range of unforeseeable activities that Mehrotra describes in his discussion of the Victorian Arcades in Mumbai.

In terms of the individual housing units, they consist of a basic living structure comprising the kitchen and one full bathroom, the rest is an open plan. As explained above, these are the parts of the house which require the most technical support: structure, plumbing and electrical arrangements. Residents can carry out further developments within the confines of this basic structure at a very low cost without technical assistance. Sometimes gaps are left between housing units, so that houses can expand into the void. When higher densities are required, dwellers receive an empty multi-storey volume with the kitchen and a bathroom. This basic structure allows residents to build extra floors inside the original empty volume which also contains mechanical and electrical installations. Though incomplete, this basic unit is larger than the standard social or low-income house commonly offered to poor people in Chile. In other words, the small subsidies are used to build a large, stable and properly serviced shell which, though initially empty, when completed by the owners will almost reach the size of an average middle-income house. In all cases dwellers are free to customise their dwellings according to their own needs and variable economic capacities – for there is little labour stability. Further subsidies can be obtained but these are, indeed, rare.

… in this case architecture is not conceived as the 'spectacle of the static city', but as a celebration of its performative (or kinetic) temporality.

When the houses have developed to occupy all vacant spaces (interior and exterior), the site looks entirely different from the way it did prior to occupation. This results in vibrant urban landscapes that reveal the great heterogeneity of Chilean peoples. In other words, the projects designed by ELEMENTAL are an outlet for the expression of cultural difference; a space where diverse sociocultural groups can perform their differences and negotiate them with other dwellers on a continuous basis – not always harmoniously. Paraphrasing Mehrotra, in this case architecture is not conceived as the 'spectacle of the static city', but as a celebration of its performative (or kinetic) temporality.

To conceive of architecture in this way or, rather, to design buildings as ELEMENTAL does, presents various architectural challenges. The advancement of users to the position of producers of architecture, for example, removes the architect from its hegemonic place as author. It will no longer be possible to link buildings to the single architect, or architectural practice, who designed them. The term 'translatable architecture' comes to mind, linking Benjamin's notion of translation and Bhabha's idea of cultural translatability (see Chapter 2). We might say that ELEMENTAL presents a situation where the architect and the building touch only slightly, tangentially, at one point in history after which the building follows its own path of historical becoming in the hands of the people. As a result, the authority granted to architecture, and the 'author-architect', is disturbed by the fact that buildings are constantly being re-signified by users. This embodies the conflict that Bhabha illustrates in his discussion about the pedagogical and performative. Borrowing from Bhabha, it can be argued that to become intelligible, the processes of design employed by ELEMENTAL requires a 'cultural temporality that is both disjunctive and capable of articulating forms of activity which are both at once architectural and not'. More importantly, this mode of practice decidedly creates a space of architectural productivity in the interstices between pedagogical architectural narratives (the principles of architectural composition, structures, etc.) and the performativity of habitation. It is not that people's performances ultimately obliterate architecture altogether, but that the basis of its authority is unsettled.

Indeed, due to their 'performative', 'kinetic' and, even, 'translatable' characteristics the buildings designed by ELEMENTAL refuse rapid inclusion in the history of architecture. As discussed in Chapter 3, architectural history registers completed buildings mainly on the basis of their formal characteristics so that they can be classified in relation to western historicity – which links architectural production around the world with Europe's past. Since the form and image of the buildings here examined are changing constantly, it is difficult to classify them formally or to locate them within a particular historical period. What is more, as cultural (rather than architectural) hybrids the buildings in question do not fit national associations: they are not Chilean colonial nor Chilean modernism, nor do they fit the description vernacular architecture (Chilean vernacular). Instead they express the tense coexistence of different cultures and forms of dwelling, and so they exceed any simplified version of Chileaness – they represent the hybridity of Chilean cultures in the way Bhabha describes, not simply through their form and image. The performative, or kinetic, temporality of the architecture produced by ELEMENTAL is accentuated deliberately in the exhibitions of their work. Every time they exhibit, ELEMENTAL prefers to show images of the 'present state' of the buildings, rather than photographs of the structures when they were just finished. Since the buildings are constantly being altered by users, they look different every time they are exhibited – once again the idea of translation comes to mind, for it could be argued that every time the same building is exhibited, it 'attains to its ever-renewed latest and most abundant flowering'.

Two aspects ought to be underlined in this brief discussion about the performative temporality of the work of ELEMENTAL. First is that their approach to social housing introduces an important political variable to the design of buildings, one in which users are conceived of as producers of their own habitable space. Second is the consequent dissociation between the author (architect) and the building, as well as between the building and its image, which can no longer be considered immutable. While this approach certainly corresponds more accurately with the swiftly changing socio-political circumstances that surround the lives of the Chilean poor – and the poor in other countries in the developing world, as Yeoh and Mehrotra demonstrate – it also presents a scholarly challenge because existing methods of architectural historicisation are unsuitable to deal with the exacerbated dynamism inherent in their buildings. Hence, in the same way that architects have had to develop alternative methods of design to respond to the specific circumstances of the poor, scholars are faced with the need to produce suitable methods to record historically and theorise hybrid and performative (kinetic) architectures.

CHAPTER 7

Conclusion

In previous chapters, we have focused largely on the specificities of the colonial relation and, by implication, on the period of colonialism. As mentioned in the Introduction, the period of colonialism may seem distant to a generation born in the last 30 years of the twentieth century, and others thereafter. However, the historical proximity of this period is such that many people can still give first-hand accounts of their experiences of colonialism; both as subjects and as rulers. The proximity of that seemingly distant era explains why many of the strategies used to construct and exercise colonial authority are still employed today, although under different guises. Indeed, through the analysis of architectural case studies this book demonstrates that architecture, as a pedagogical narrative, is in many ways complicit with the continuation of colonialist methods of representation which construct the non-western as inferior. This is achieved, amongst other ways, through the derogatory inscription of non-western architectures in the hegemonic, and conspicuously singular, western architectural historicity.

Chapter 2 offers a simple explanation of the methods used to construct and maintain authority in relations of colonialism, and provides some historical examples. More importantly, this chapter introduces the terminology of postcolonial discourse and the principles of its critique. Chapter 3 provides a detailed explanation of the psychoanalytic concept of ambivalence, a key term in Bhabha's critique of authoritarian discourse. This concept is used in order to reveal the contradictions inherent in colonialist narratives and, so, it helps to undermine the arguments used which justify both colonisation and contemporary structures of domination (economic, cultural and political). Chapter 3 also illustrates how the historical inscription of non-western architectures continues to occur according to the ambivalent process of inclusion and exclusion which lies at the centre of Bhabha's critique. Chapters 4 and 5 show that colonialism did not result in a complete merger between colonised and coloniser, but in the emergence of a complex process of cultural interaction between multiple parties; a process that Bhabha names hybridisation. Rather than harmonious, hybridisation is a conflictive process in which cultural differences are continually negotiated and contested, a process which reveals the inequalities of both colonialism and contemporary cultural interaction. The various forms of cultural rearticulation examined in this book did not stop when direct colonial control ended – when the colonies became independent. Cultural interaction continued and, in fact, increased, after the end of the colonialism. That is why Chapter 6 focuses on the conditions of life today in the context of the modern nation; an artificial construct that strives towards stability and homogeneity but which, in the process, obscures the existence of cultural difference.

In explaining Bhabha's terminology we have moved gradually from the height of colonialism in the nineteenth century to the present day. We have reviewed primarily two moments in the history of colonialism. First, the moment of direct intervention when the coloniser attempted to construct itself as an authority, while simultaneously constructing the colonised as a subject; as epitomised in the now infamous 'Minute on Indian Education' written by Thomas Macaulay in 1835. The second moment corresponds with the still-felt effects of colonialism today. Indeed, Bhabha's discussion on culture entails that contemporary international relations are still largely determined by the social and political structures built during the colonial era. That is why the remit of postcolonial discourse, and the scope of Bhabha's critique, is not limited to the study of colonial relations. Postcolonial discourse is also concerned with contemporary cultural relations, as well as with issues regarding cultural representation and the construction of cultural identities in our increasingly globalised era.

The arguments put forward by Homi K. Bhabha, his incisive critique of authoritarian discourses and his interest in foregrounding the role of 'people' – minorities and the subaltern – in the continuous construction of culture (historically and in the present) appear to be useful in order to advance a critique of architecture; not only the historicisation and theorisation of architecture but, also, its professional practice. As this volume demonstrates, non-western and minoritarian architectures can only emerge, become visible (and recognisable), in the terms of European and North American scholarly narratives. The arguments, terminology and the architectural case studies examined in this volume show that it is possible and, indeed, necessary, to question the hegemonic normality assigned to those architectural postures which perpetuate the uneven and differential historical inscription and theorisation of non-western and minoritarian architectures. While a postcolonial critique of architecture following Bhabha's model may not overturn completely the authority of the west, it would certainly be useful to reveal its intrinsic contradictions and, so, to destabilise the hegemony of architectural discourse; to render it questionable and to distort its rules of recognition.

Works Cited

Abel, C. (1997) *Architecture and Identity: Responses to Cultural and Technological Change*. Oxford: Architectural Press.

Anderson, B. (1983) *Imagined Communities: Reflections on the Origins and Spread of Nationalism*. London: Verso.

Benjamin, W. (1968) 'The Task of the Translator', in Hannah Arendt, *Illuminations*. Harry Zohn (trans.). New York: Schocken Books, 70–82.

Bhabha, H. (1990a) *Nation and Narration*. London and New York: Routledge.

—— (1990b) 'Third Space', in J. Rutherford (ed.), *Identity: Community, Culture and Difference*. London: Lawrence & Wishart, 207–21.

—— (1993) 'Cultures in Between'. *Artforum* 32.1, 167–214.

—— (1994) *The Location of Culture*. London and New York: Routledge.

—— (2004) *The Location of Culture*. London and New York: Routledge Classics.

—— (2007) 'Architecture and Thought', in *Aga Khan Award for Architecture: Tenth Cycle*. AKAA Publications.

Conrad, J. (1995) *Heart of Darkness*. London: Penguin Books.

Curtis, W. J. R. (2000) *Modern Architecture since 1900*. London: Phaidon.

Fanon, F. (2008) *Black Skins, White Masks*. Oxford: Blackwell.

Freud, S. (2002) *Civilization and Its Discontents*. London: Penguin Classics.

Grahn, L. R. (1995) 'Guajiro Culture and Capuchin Evangelization: Missionary Failure on the Riohacha Fontier', in E. Langer and R. Jackson (eds), *The New Latin American Mission History*. Lincoln: University of Nebraska Press, 130–56.

Hernández, J. (1998) *Martin Fierro*. Madrid: Ediciones Cátedra.

Hobsbawm, E. J. (1987) *The Age of Empire, 1875–1914*. New York: Pantheon Books.

Homer, S. (2005) *Jacques Lacan*. London and New York: Routledge.

Ikas, K. and G. Wagner (eds) (2009) *Communicating in the Third Space*. London and New York: Routledge.

Jacobs, J. M. (1996) *Edge of Empire: Postcolonialism and the City*. London and New York: Routledge.

Lacan, J. (2006 [1996]). *Écrits*. B. Fink (trans.). New York and London: W. W. Norton & Co.

Lefebvre, H. (2003 [1991]) *The Production of Space*. Oxford: Blackwell Publishing.

Macaulay, T. (1835) 'Minute on the 2nd of February 1835', in *Speeches by Lord Macaulay, with His Minute on Indian Education*. With Introduction by G. M. Young. London: Oxford University Press, 340–50.

Mehrotra, R. (2010) 'Foreword', in F. Hernández, P. Kellett and L. Allen (eds), *Rethinking the Informal City: Critical Perspectives from Latin America*. Oxford and New York: Berghahn Books, xi–xiv.

Morton, P. A. (2003) *Hybrid Modernities: Architecture and Representation at the 1931 Paris Colonial Exposition*. Cambridge, MA: MIT Press.

Niranjana T. (1992) *Siting Translation: History, Post-Structuralism and the Colonial Context*. Berkeley: University of California Press.

Pratt, M. L. (1992) *Imperial Eyes: Travel Writing and Transculturation*. London and New York: Routledge.

Said, E. (2003) *Orientalism: Western Conceptions of the Orient*. London: Penguin Classics.

Soja, E. (1996) *Thirdspace: Journeys to Los Angeles and other Real-and-Imaginary Places*. Malden and Oxford: Blackwell Publishing.

Yeoh, B. S. A. (1996) *Contesting Space: Power Relations in the Urban Built Environment in Colonial Singapore*. Oxford: Oxford University Press.

Young, R. (1990) *White Mythologies: Writing History and the West*. London and New York: Routledge.

Further Reading

There are various books that explain the work of Homi K. Bhabha in the context of postcolonial theory and literary studies. These books are helpful to understand Bhabha's terminology: hybridity, mimicry, Third Space, etc. For example *Homi K. Bhabha* by David Huddart (2006) and *Homi K. Bhabha* by Eleanor Byrne (2008). These two books also provide a complete bibliography of Bhabha's essays and articles which are scattered amongst many books and journals.

Bhabha's work has also been discussed critically by various postcolonial theorists. For example: *White Mythologies: Writing History and the West* by Robert Young (1990) and *The Empire Writes Back: Theory and Practice in Postcolonial Literatures* by Bill Ashcroft, Gareth Griffiths and Helen Tiffin (1989). These books discuss the work of Bhabha and other critics. The two books present a broad panorama of early discussions in postcolonial theory. Bill Ashcroft, Gareth Griffiths and Helen Tiffin also wrote a book entitled *Post-Colonial Studies: Key Concepts* (1998), which explains some of the terms commonly used in postcolonial theory generally.

There are many books that use Bhabha's theories in the context of architecture, so it is difficult to single out one as the most useful. However, the ArchiText series published by Routledge contains various volumes that could help to understand how Bhabha's theories have been used in architectural studies. This is a list of some books in that series which mention the work of Bhabha: *Behind the Postcolonial: Architecture, Urban Space and Political Cultures in Indonesia* (2000) by Abidin Kusno; *Drifting: Architecture and Migrancy* (2003) edited by Stephen Cairns; *Spaces of Global Cultures: Architecture, Urbanism, Identity* (2004) by Anthony King; *Indigenous Modernities: Negotiating Architecture and Urbanism* (2005) by Jyoti Hosagrahar; *Colonial Modernities: Building, Dwelling* and *Architecture in British India and Ceylon* (2007) edited by Peter Scriver and Vikramaditya Prakash.

Bhabha, in collaboration with Carol A. Breckenbridge and Sheldon Pollock, edited a book entitled *Cosmopolitanism* (2002) which addresses many aspects of cities, buildings and urban spaces from a postcolonial perspective. Although none of the contributors to this volume is an architect, the book is an example of the way in which deconstructive methods of analysis can be used in order to dismantle the narratives that underpin the construction of metropolitan space.

There are two other authors whose work occupies a prominent position in postcolonial studies: Frantz Fanon and Edward Said. Fanon wrote extensively, but two books stand out: *Black Skins, White Masks* (2008) and *The Wretched of the Earth* (2001). Edward Said was also a prolific writer, however one book made a profound mark on contemporary academia: *Orientalism: Western Conceptions of the Other* (2003).

Index

Abel, Chris 78–82, 87
Aga Khan Award for Architecture 18
The Age of Empire 1875–1914 (Hobsbawm) 104
Algeria 2
'almost the same but not quite' 64–5
ambivalence 51; in architectural history writing 51; of colonial discourse 41, 43–8, 53; context of Bhabha's 39–41; during the Oedipus phase 42; in Fanon 40–1; Freudian perspective 42–3; in Grahn's account of the Guajiro Indians 48–9; and the nation 45; in psychoanalysis 42–3; and the Third Space 93; usefulness in producing a critique of architecture 50; Young on Bhabha's discussion of 48
Anderson, Benedict 92, 99, 104, 107, 111
Aravena, Alejandro 123–4
architectural history: ambivalence of *see* ambivalence; Bhabha on the writing of 57; hegemonic narratives in 52; role of the book 52
architectural hybridity: Abel's account 78–82; in the Paris Exposition 83–6 *see also* hybridity/hybridisation

architectural identity 20, 81
architectural informality 99
architectural narratives 8, 51–2, 127
architectural production 1, 8, 18, 20–1, 51, 53–5, 100, 128
architectural theory 98
architecture: colonial *see* colonial architecture; European 8, 17, 43, 51, 77–81; indigenous *see* indigenous architecture; non-western *see* non-western architectures; Western 55
Ashcroft, Bill 13, 76
Azaldúa, Gloria 96

Barthes, Roland 10
Benjamin, Walter 10, 25–30, 32, 34–5, 37–8
Bhabha: academic positions 9; arrival at Oxford 2, 6; biographical account 8–9; birth 5; critical methods of reading 13; cultural and literary interests 6; interest in architecture 18–20; life and journey 3–5
the Bible 61–2, 67, 70
Black Skins, White Masks (Fanon) 40
Bochica 33, 35
Bombay (Mumbai), Bhabha's description 5; *see also* Mumbai

Braveheart (Gibson) 108–9, 113
Brazil 17, 54, 56
Brick Lane, East London 111
building(s): architectural 'meaning' inherent in 56; Bhabha's 'reading' 20–1

Chile 123–5
Chinese-built 'shophouse,' Abel's study 79
Christianity 25, 33, 48, 59, 61, 68
cities 1, 5, 8, 16–18, 21, 73, 96–9, 117–24
civilisation, Christianity as sign of 25
Civilisation and its Discontents (Freud) 42
civilising mission, Macaulay on the purpose of the 64
Colombia 33
colonial architecture, Georgetown, Penang **78**
colonial cities: educational function 16–17; geometrical layout and the desired social order 118; participants in the consolidation of 118; Yeoh's criticism of the dual approach to the study of 117–20
colonial mimicry 64–5, 81
colonial relations: Hobsbawm on racism and 106; and hybridity 76; implicit hierarchy 25
colonial stereotyping 32
colonial translation, function 34
colonialism: ambivalence in the discourse of 41; Bhabha's interest in the effects of 15; Bhabha's psychoanalytic re-imagining 12–13; concept analysis 1; forms of translation involved in 24; historical proximity of 2; as 'translational' phenomenon 30
contemporary cities, performative temporality of 120, 120–2
Contesting Space: Power Relations in the Urban Built Environment in Colonial Singapore (Yeoh) 117
criollo 108
cultural difference: concept analysis 11, 99, 100–4; cultural diversity vs 100–1; minority positions and the notion of 103; result of Bhabha's dismantling of homogeneity on 67
cultural diversity, and the *musée imaginaire* 101–2
cultural identity, sources of 110
cultural productivity: hybridity as sign of 49, 58–9, 66–7, 72; location 6
cultural superiority 25, 65–6
cultural translation 31–2, 37, 39
Curtis, William 17, 54–6

de Man, Paul 38
deconstruction 8, 10, 30, 46
Derrida, Jacques 10, 38
discrimination: Bhabha on 65; constructing authority through 43–4; hybridity/hybridisation and 74; as means to bring betterment 63
'DissemiNation: Time, Narrative and

the Margins of the Modern Nation'
(Bhabha) 108
domination, architects' complicity
with colonial and other discourses
of 17
Doshi, Balkrishna 17
doubling 12–13, 45, 47, 65

Edge of Empire: Postcolonialism and the City (Jacobs) 73
ELEMENTAL, Chile 123–8
'end of empire' 2
English education, Macaulay on the function of 33
Enlightenment 63–4, 104
Equatorial Guinea 2
European architecture 8, 17, 43, 51, 77–81

Fanon, Frantz 1, 40–1
French architecture, hybridity in 83–7
Freud, Sigmund 12, 42–3, 45–6, 50

Gacaca Courts, as Third Space 93
Gauchos 108
gender, role in the formation of identity 46
Georgetown, Penang – British colonial house **78**
Grahn, Lance 48–9
'Guajiro Culture and Capuchin Evangelisation: Missionary Failure on the Riohacha Frontier' (Grahn) 48
Guajiro Indians 48–9, 68

Hall, Stuart 40
Heart of Darkness (Conrad) 94
Hegel, Georg W. F. 110
heterogeneity 38, 59, 66, 110, 114–15, 126
Hinduism 61
historicism 110
historicity 100, 113
history of architecture, European authority 17
Hobsbawm, Eric 99, 104–7
homogeneity/homogenisation: of colonised subjects 34; the completeness of identity and 12; concept analysis 112–13; of the nation 110, 115; result of Bhabha's dismantling on cultural difference 67
'How Newness Enters the World' (Bhabha) 28
Humboldt, Alexander 52–4
hybrid architecture 77, 80, 87; *see also* hybridity/hybridisation
hybridity/hybridisation: adverse implications 58–9; and ambivalence 41, 44; appeal of the notion of 58; Bhabha on 66, 69; colonial relations and 76; concept analysis 58, 60–73; critiques of 73–7; cultural 11, 58; and cultural productivity 72; and discrimination 74; embodiment of Babha's concept of 64; of European architectural styles 81; as form in architecture 77–82; Grahn's account of Guajiro resistance as example of 49; impact of permanence of 11;

hybridity/hybridisation *(continued)*
Morton's perspective 83–5, 87; of the mule 59; in the Paris Exposition 83–6; postcolonial 76; as result of cultural translation 37; semantic implications 77; as sign of cultural productivity 59; theoretical effects 58; theoretical optimisation 77; and the Third Space 89–90, 92; as threat 86; Young on Bhabha's use of 74–5

identity/identities: Bhabha's representation 5; gender's role in the formation of 46; homogeneity and the completeness of 12; language and 46; linguistic perspective 46; plurality of 12, 40–1, 59; psychoanalytic perspective 12, 40–1; relative construction 9

Imperial Eyes: Travel Writing and Transculturation (Pratt) 52

imperialism 40, 49

India 2

indigenous architecture 8; Abel's study of Malaysian 78–81; colonial perspectives 8, 53–4; Pratt on Central American 52

Jacobs, Jane 73–4, 76

Japan 54, 56

kinetic space 121–2

Kristeva, Julia 12

Lacan, Jacques 12, 45–7, 50, 90–1

languages, Benjamin on the dynamism of 26–7

Le Corbusier 2, 17, 53, 55

Liberty Leading the People (Delacroix) 109, 113

liminality 89–90, 95, 114–16

literary translation, Benjamin's theory of 10

The Location of Culture (Bhabha) 3, 8–9, 57, 60, 64, 108

Loomba, Ania 76–7

Macaulay, T. 33, 62–4, 66, 92, 102, 110–11

Macaulay's Minute 62–4, 66, 92

Malay house **78**, 80

Malaysia 17, 78–81

Malaysian architecture, Abel's study 78–81

Martin Fierro (Hernández) 108, 113

materials 20, 73, 77, 82, 84, 87, 120

Mehrotra, Rahul 120–1, 123, 125–6, 128

Mexico 54, 56

Mies van der Rohe, Ludwig 53

migrant minorities, Bhabha on 36–7

mimicry: ambivalence and 41, 44; Bhabha's definition 64; colonial 64–5, 81; concept analysis 65; Young's criticism of Bhabha's use of 74

minoritarian architecture 17, 23, 132

minority peoples: Bhabha's interest in the cultural products of marginalised 7; cultural difference

and the agency of 100–4; and the
Third Space 96–8
'Minute on Indian Education'
(Macaulay) 62–4, 66, 92
missionaries 33–4, 36, 48
modern architecture 8, 17, 53–7, 100;
genealogy 53–5
Modern Architecture since 1900
(Curtis) 54–6
modern art, market domination 16
Morton, Patricia 82–8
Mozambique 2
Muisca 33, 35–6
multiculturalism: vs cultural difference
103; cultural stratification rhetoric
11; images of 112; and racism 102
Mumbai: Bhabha's description 5;
Victorian Arcades 121, **122**, 125
musée imaginaire 101–2
Muslim world, Bhabha's definition
18–19

narcissistic demand 41, 66
the nation: Anderson's definition
107; Bhabha's critique 108–16;
contradiction between the newness
of and its alleged antiquity 111–12;
'the historical certainty' evoked by
the concept of 110; Hobsbawm on
105–7; pedagogical and
performative temporalities 113;
and some ideas on nationalism
104–8
Nation and Narration (Bhabha) 108
the nation-space, liminality 115–16

the nation-state, historical emergence
104
nationalism, Hobsbawm and
Anderson's observations 104
neocolonialism 40
Niemeyer, Oscar 17
Niranjana, Tejaswini 35, 38
non-western architectures: basis of
the historicisation of 20, 51, 57;
binary methods of analysis 117;
critique of theoretical perspectives
7, 16–17; Curtis' terminology 54,
56–7; discriminatory rhetoric 52;
see also indigenous architecture

Oedipus complex 42
'Of Mimicry and Man' (Bhabha) 64–5
Orientalism 83
orthogonal grid 17, 25
Other: dismantling of the
straightforward dichotomy between
self and the 41; mimicry as desire
for a reformed, recognisable 64

Pakistan 2
Palestine 56
Paris 82, 84–6
Parry, Benita 76–7
Parsis, background and philosophy
4–5
pavilions 83–4
the performative: architecture and 123,
123–9; comparability of kinetic
space with 121–2; concept analysis
114; temporal perspective 113

Pieterse, Jan Nederveen 77
the poor 11, 15, 97–8, 103–4, 121, 124, 128–9
post-structuralism 9–10, 12, 35, 39
postcolonial architecture discourse, Bhabha's domination of 1
postcolonial criticism, Bhabha on 14
postcolonial hybridity, critical efficacy of the concept of 76–7
postcolonial reading, Ashcroft on 13
postcolonial theory, concept analysis 14
postcolonial translation 37–8
poverty: the perpetuation of 16; and the Third Space 97–8
Pratt, Mary Louise 52–3
The Production of Space (Lefebvre) 96
psychoanalysis: ambivalence of 42; and the completeness of identity and homogeneity 12; critiques of Bhabha's appropriation 49–50, 75; influence of 12, 39; Lacanian 45–7, 91
psychoanalytic ambivalence 39, 47, 50, 73

racial discrimination, Hobsbawm on 106
racism, multiculturalism and 102
Republic of Congo 2
resistance 38, 47, 68, 74, 87, 119

Said, Edward 1, 40, 83, 96
Shohat, Ella 77
'Signs Taken for Wonders' (Bhabha) 60, 65

Singapore 117, 119
Siting Translation: History, Post-Structuralism and the Colonial Context (Niranjana) 35
slums 8, 18, 97–9
Soja, Edward 96–7
South Africa 54, 56
South America 19, 48
Spivak, Gayatri 1, 40, 96
squatter settlements 8, 18, 98–9

Tableaux Parisiens (Baudelaire) 26
'Task of the Translator' (Benjamin) 26
'The Death of the Author' (Barthes) 10
'the English book' (the Bible) 60–2, 64, 67–8, 70–1
The Tempest (Shakespeare) 26, 29
Third Space: ambivalence and the 91–3; architecture and the 96–8; concept analysis 21, 89–90; courtroom comparisons 93–4; in *Heart of Darkness* plot 94; hybridity/ hybridisation and the 89, 92; liminality of the 89, 95; minority peoples and the 96–8; Soja on the 96–7; spatialising the 93–5; squatter settlements as 97–8; theorising the 90–3; Truth Commissions as 93
translation: Benjamin on 26–8; Bhabha's use of as a critical term 25; bread example 28–9; colonial 34–5, 37; concept analysis 24;

cultural 31–2, 37, 39; and the elimination of differences 34; function of colonial 34; of the history of colonised subjects 33; political dimension 36; postcolonial 37–8; re-creation vs 'copy' 30; as tool of colonial authority 31

Truth Commissions, as Third Space 93

uprising 63

Victorian Arcades, Mumbai 121, **122**, 125

Wallace, William 108
Western architecture 55
Wright, Frank Lloyd 53–4

Yeang, Ken 17
Yeoh, Brenda 117–20, 123, 128
Young, Robert 48, 74–7

eBooks – at www.eBookstore.tandf.co.uk

A library at your fingertips!

eBooks are electronic versions of printed books. You can store them on your PC/laptop or browse them online.

They have advantages for anyone needing rapid access to a wide variety of published, copyright information.

eBooks can help your research by enabling you to bookmark chapters, annotate text and use instant searches to find specific words or phrases. Several eBook files would fit on even a small laptop or PDA.

NEW: Save money by eSubscribing: cheap, online access to any eBook for as long as you need it.

Annual subscription packages

We now offer special low-cost bulk subscriptions to packages of eBooks in certain subject areas. These are available to libraries or to individuals.

For more information please contact webmaster.ebooks@tandf.co.uk

We're continually developing the eBook concept, so keep up to date by visiting the website.

www.eBookstore.tandf.co.uk